The RCS Handbook

The RCS Handbook

Tools for Real-Time Control Systems Software Development

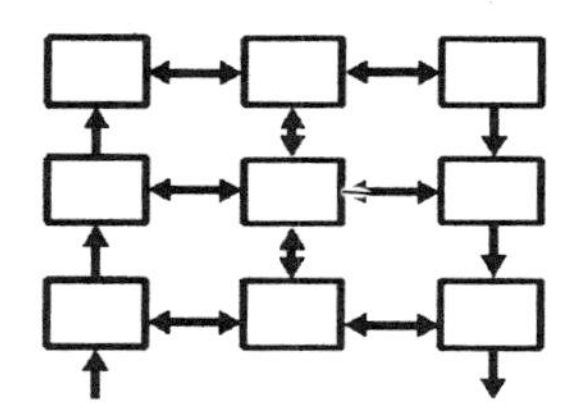

Veysel Gazi, Mathew L. Moore, Kevin M. Passino
Department of Electrical Engineering
The Ohio State University

William P. Shackleford, Frederick M. Proctor, James S. Albus
Intelligent System Division
National Institute of Standards and Technology

A WILEY-INTERSCIENCE PUBLICATION
JOHN WILEY & SONS, INC.
New York • Chichester • Weinheim • Brisbane • Singapore • Toronto

This text is printed on acid-free paper. ♾

Published simultaneously in Canada.

For ordering and customer service, call 1-800-CALL-WILEY.

Library of Congress Cataloging-in-Publication Data:

Gazi, Veysel.
The RCS handbook : tools for real-time control systems software development / by Veysel Gazi . . . [et al.].
p. cm.
Includes index.
ISBN 0-471-43565-1 (cloth : alk. paper)
1. Real-time control. 2. Real-time data processing. I. Title.

TJ217.7 .G39 2001
629.8'.95433—dc21 2001017667

Printed in the United States of America.

10 9 8 7 6 5 4 3 2 1

To my parents (V. G.)

To my family (M. L. M.)

To Annie, Carina, and Juliana (K. M. P.)

To David, Rob, and Linda (W. P. S.)

To Amy, Alex, and Becky (F. M. P.)

To Cheryl (J. S. A.)

Contents

Preface

Control systems are becoming increasingly complex and hence there is a growing need for software tools for design and implementation that can ease the burden of deploying such systems. For instance, in robotic, manufacturing, and process control systems, sensors, actuators, and components of the controller responsible for different tasks are physically distributed across the process, and are possibly implemented on different computer platforms (e.g., a force-feedback controller may be implemented with a microprocessor placed near the gripper of a robot, yet robot path planning functions may be implemented on a different computer). This creates two needs. First, we need a method to communicate between these possibly distant controllers, each of which can be implemented with a different technology. Second, for many applications we most often need a controller to supervise and coordinate the activities of the multitude of algorithms so that the overall control objectives can be met. This creates a hierarchy of distributed control algorithms, where each algorithm may reside on a different computer. For instance, in a robotic system there is typically a need for a centralized controller that can coordinate the actions of many subordinate controllers that are responsible for force feedback, endpoint positioning, and trajectory planning, to name a few.

This book provides a description of a set of software tools, the National Institute of Standards and Technology (NIST) real-time control systems (RCS) library, which has proven to be very useful in the development of hierarchical and distributed real-time control systems for a wide variety of applications.

Overview of the Book

The book is broken into three parts. In Part I we provide a tutorial introduction to RCS. In Part II we provide an RCS handbook for the RCS applications engineer. Part III provides some appendices that provide some useful background information.

Part I begins with Chapter 1, where we provide a brief overview of the RCS library and this book. In Chapter 2 we provide a "quick start" tutorial introduction to the RCS software. To do this we use a simple laboratory application and show every step in the development of a simple RCS application. In Chapter 3 we describe a Reference Model Architecture (RMA) that provides a framework to structure complex controllers, and a variety of applications where the RMA has been implemented successfully via RCS. The intent of Chapter 3 is simply to give an idea of the strengths of the RCS software by means of examples of real-world applications. It also serves as an introduction to [7], a book in which, the underlying theory, and the use of the RMA and RCS for sophisticated applications, is discussed in detail.

Part II starts with Chapter 4, where we introduce a sample application, an automated highway system, that we will use as a theme problem throughout the remainder of the book to illustrate many of the ideas and methods of the RCS software. Next, in Chapter 5 we discuss some of the core ideas of the RCS library when we cover programming in NML (*Neutral Message Language*). In particular, we cover NML message classes, NML communication channels, how to read and write NML data, error handling, how to spawn and kill NML servers, and some user command utilities. In Chapter 6 we discuss the RCS control module, and in Chapter 7 we cover how to write NML configuration files (which establish the layout and interconnections of the hierarchy). In Chapter 8 we discuss other classes and functions of the RCS library, including the timer, semaphore, linked list, print, and window functions. In Chapter 9 we discuss the *diagnostics tool*, which is used to monitor (and guide, if needed) the operation of a hierarchical distributed controller that is implemented with the RCS library. This tool can be very useful in development and debugging. In Chapter 10 we discuss an automatic code generation feature of the RCS library and the *design tool*. The design tool provides a graphical method to construct hierarchical distributed control system hierarchies and hence relieves the user of low-level coding details. In Chapter 11 we discuss the *architecture file* that is used in both the RCS design and diagnostics tools.

Part III holds the appendices of the book. Appendix A provides a brief introduction to C++ for those already familiar with C. Appendix B discusses compilers and makefiles, and Appendix C overviews some key operating system concepts and computer network communication protocols. Appendix D discusses RCS version functions, and Appendix E provides a table of platforms tested.

How to Get the Code for the NIST RCS Library

One of the especially attractive features of the NIST RCS library is that it is *free* of charge and you can download it from

`ftp://ftp.isd.mel.nist.gov/pub/rcslib`

If you have general questions about RCS, you should contact the NIST team for more details via

`http://www.isd.mel.nist.gov/projects/rcslib`

Intended Audience

Control engineers from industry will find the book useful as a handbook for use of the RCS library to develop complex control systems. Researchers will find it useful as a complete treatment of software tools for hierarchical and distributed control systems development. Moreover, researchers in the area of *intelligent control* will find it useful as a software user's manual for implementation of the *Reference Model Architecture* for complex applications as they are outlined in [7].

Universities will find it useful as a book to be used in conjunction with laboratories that use the software, or classes that discuss issues surrounding the implementation of complex distributed real-time control systems. Indeed, a lecture/laboratory course has been taught from this book at The Ohio State University (OSU). For this, there are lectures on the main RCS topics and the students use the RCS library to implement two different controllers. Part I of this book has also served as the background material for a two-week laboratory project in RCS development for the tank system described in Chapter 3. Hence, Part I can serve as the basis for a two-week project on the implementation of distributed controllers in an existing university control laboratory course. RCS is also being used as a tool in senior design projects at OSU.

Acknowledgments

First, it is important to clarify how this book came to be. To begin with, the NIST team spent years of development work on the RCS library and fine-tuned the details by using the library to implement a variety of complex control systems. More recently, under a grant from NIST, OSU developed a lecture/laboratory course on the use of the RCS library, the two-week "quick start" RCS project on the tank (Part I), and took the lead in the writing project for this book. Everything in this book relies on earlier work done at NIST, and the NIST team wrote the initial versions of much of this book (e.g., for Part II). OSU added many examples, particularly the process control problem in Part I and the automated highway system theme problem for Part II. In terms of level of contributions, the ordering of the names within the OSU and NIST teams is accurate; however, listing the OSU team before the NIST team does not imply that any one OSU team member contributed more than any other member of the NIST team.

The development of such a complex software package requires the cooperation of many people and it is our pleasure to acknowledge their contributions here. In particular, we would like to thank the following people, who had the courage and patience to be some of the early users and evaluators of the software and provided sound advice with their feedback: Marilyn Abrams (Nashman), Hoosh Abrishamian, Steven Balakirsky, Rogier Blom, David Coombs, Joe Falco, Jim Gilsinn, Mark Del Giorno, Tsai Hong, John Horst, Hui-Min Huang, Tom Kramer, Alberto Lacaze, Steven Legowik, Todd Litwin, Elena Messina,

John Michaloski, Karl Murphy, Rick Quintero, Bill Rippey, Bob Russell, Minbo Shim, Keith Stouffer, Sandor Szabo, Tsung-Ming Tsai, Tom Wheatley, and Bill Yoshimi.

We would also like to thank Nat Frampton, Anthony Barberra, and Advanced Technology and Research Corp. for providing us access to some of their earlier RCS tools and graciously answering questions on their methodology. At OSU we would like to thank Andy Bushong and Xiaoqiu Li for their feedback on the use of the package in a university setting, and Raúl Ordóñez and Murat Zeren for their valuable discussions and suggestions.

We would like to thank the publishers at Wiley, and in particular George Telecki, for all their support in publishing this book. Finally, we would like to thank our families for their support during the development and writing of this book.

Commercial Product Disclaimer

DISCLAIMER: Commercial equipment and materials are identified in order to adequately specify certain procedures. In no case does such identification imply recommendation or endorsement by the National Institute of Standards and Technology, nor does it imply that the materials or equipment identified are necessarily the best available for the purpose.

Veysel Gazi, Matt Moore, Kevin Passino
Columbus, Ohio

Will Shackleford, Fred Proctor, Jim Albus
Gaithersburg, Maryland

The RCS Handbook

Part I

RCS Tutorial

Chapter 1

Introduction

Increasing consumer and governmental demands for automation, fueled by significant advances in computing and communications technologies, have resulted in the creation of hierarchical and distributed controllers that achieve high levels of automation (e.g., via automating tasks in robotic, manufacturing, and vehicular applications that were traditionally performed by humans). The real-time control systems (RCS) library developed by the Intelligent Systems Division of the National Institute of Standards and Technology (NIST) is a software tool that can be used to facilitate the design and implementation of complex hierarchical and distributed control systems. This book is a user's manual for the RCS software tools.

In the past few decades various techniques, based on, for example, lead-lag compensation, proportional-integral-derivative (PID) control, optimal control, feedback linearization, sliding mode control, adaptive control, robust control, fuzzy logic, neural networks, and so on, have been developed for control of linear and nonlinear plants. However, these algorithms are typically not sufficient to achieve the levels and types of automation that are currently desired. There is often a need for more sophisticated algorithms, or "hybrid" combinations of the foregoing algorithms, which can also, for instance, predict and prevent or compensate for faults, control the "discrete event" part of the plant, and generally coordinate complex sequences of behaviors of a system. If the system is physically distributed, there is often a need for distributed controllers and hence communications between, and coordination of, the subsystem controllers to achieve overall performance objectives. Also, in many applications it is often convenient to break complex tasks down into simpler subtasks which can be implemented using low-level numerical algorithms (such as some of those listed above), provided that there is a "supervisor" to coordinate the activities so that the overall task is achieved. The need for supervision of controller functions, the distributed nature of many systems, and the need in many systems for an interface to a human user (e.g., for monitoring and specifying system goals) leads to a hierarchical structure for the controller.

An example of a hierarchical structure is shown in Figure 1.1 (this is sometimes called the *functional hierarchy* of the controller). In this diagram the boxes represent different modules of the controller, each assigned a different task (or tasks), and the lines represent communication links. The diagram is organized into *layers (levels)* where the *upper level* holds the interface to the humans, the *lower level* typically holds traditional control (and estimation) functions that interface to different part of the plant via sensors and actuators, and the *middle level* typically holds modules that coordinate the actions of the low-level algorithms and carry out the actions specified by the upper level. Viewed another way, this particular diagram has an upside-down *tree-structure* and the upper level has only one module, and the leaves are the low-level modules, which interface to various parts of the plant via the sensors and actuators.

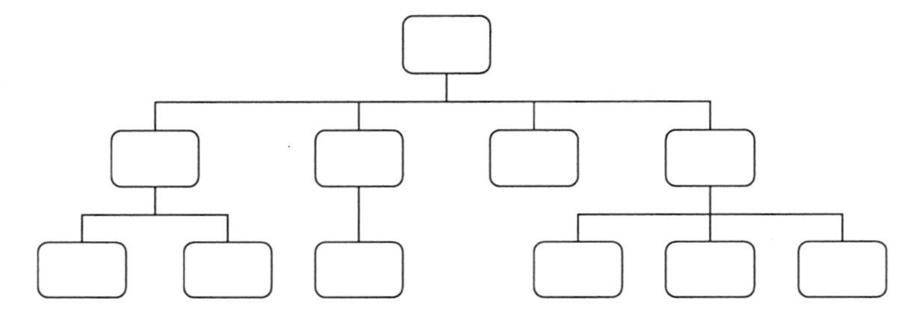

Figure 1.1: Example structure of a hierarchical and distributed control system.

The RCS library is a generalized controller software and development tool that helps to alleviate the difficulties in developing and implementing hierarchical and distributed controllers. RCS contains an inherent structure that allows for (and assists in) the decomposition of control systems into several layers and modules, as shown in Figure 1.1. RCS provides a *Communication Management System (CMS)*, which allows different routines (modules) on different platforms to communicate with each other (i.e., CMS implements the lines shown in Figure 1.1). The RCS software reduces the effort in the development of real-time control systems by providing a portable and reusable software base. The keys to the portability and standardized architecture of RCS is the use of CMS and the *Neutral Message Language (NML)*. CMS is itself a software library that contains several system-dependent operating system calls crucial to establishing communications between separate computers along a network. Several operating systems are already supported by the RCS software base, including MS-Windows, UNIX, Linux, DOS, and others. RCS provides the ability to pass information between computers running different operating systems by encoding the information in the NML format before passing it to another computer. NML provides the software base classes that allow this to happen. RCS modules communicate with each other via shared memory buffers of user-specified size. That is, sharing of information is obtained by having one process write to the buffer and other processes read the information from it. RCS provides the flexibility to establish these buffers anywhere along the network. An NML server runs in conjunction with the buffer that decodes the NML-formatted information into the native format (e.g., a format that can be used by the other processes). The NML base classes set up what information in a process is written to the buffer on a write cycle, and also provide the means of determining the type of message (e.g., what process the data came from) obtained from a buffer read. The RCS user will program mainly using the NML classes and will not deal with the CMS, so that implementation details are avoided.

The RCS library also has several tools that can be used in the development of hierarchical and distributed control systems. In particular, it provides a *diagnostics tool* that allows the user to monitor and guide the operation of a control system by providing a user interface to any module or communication channel. The *design tool* is a graphical tool that can be used to specify hierarchies such as the one in Figure 1.1, and to generate automatically much of the application-independent code needed to implement such a hierarchy.

We begin our RCS tutorial below with a "quick start," where we explain how to use RCS to implement a simple hierarchical controller. This will provide an introduction to all the key elements of the RCS library. Next, we overview how RCS has been used to implement controllers for sophisticated systems that were developed with the *Reference Model Architecture.* Part I concludes with a brief summary of RCS features. Part II provides an RCS handbook for the applications engineer, and Part III provides some background information in several appendices.

Chapter 2

Getting Started Quickly: RCS Essentials

The objective of this chapter is to provide a tutorial for the beginner RCS application developer. To do this we show each step in the design of an RCS controller for a process control experiment. To start, we introduce the system to be controlled and state the control problem. Following that we introduce the RCS design methodology and apply it to the controller development for the process control experiment. Then, we introduce some RCS library terminology and show how to generate the code for this application using the RCS design tool. After that, we present some of the code used to implement the controller to help the reader to better understand the programming issues in RCS. Finally, we present the user interface to the application that is implemented via the RCS diagnostics tool.

2.1 Process Control Experiment: The Tutorial Theme Problem

The experimental setup of the process control experiment consists of two tanks, two pumps, two temperature transducers, two level measuring devices, two mixers, and two heaters [58, 59]. The two tanks are labeled *storage tank* and *reaction tank* (see Figure 2.1). The level of the liquid in each tank is measured by a potentiometer attached to a Styrofoam float. The temperature measurements are made via temperature transducers that are mounted in the tanks. The pump that removes liquid from the reaction tank and fills in the storage tank is an AC (alternating current) pump, whereas the pump that removes liquid from the storage tank and fills in the reaction tank is a DC (direct current) pump. The AC pump, and the heaters and stirrers, can be turned on or off independently. Therefore, they compose the plant's five discrete inputs. The flow rate of the DC pump, on the other hand, can be changed by varying its supply voltage (this

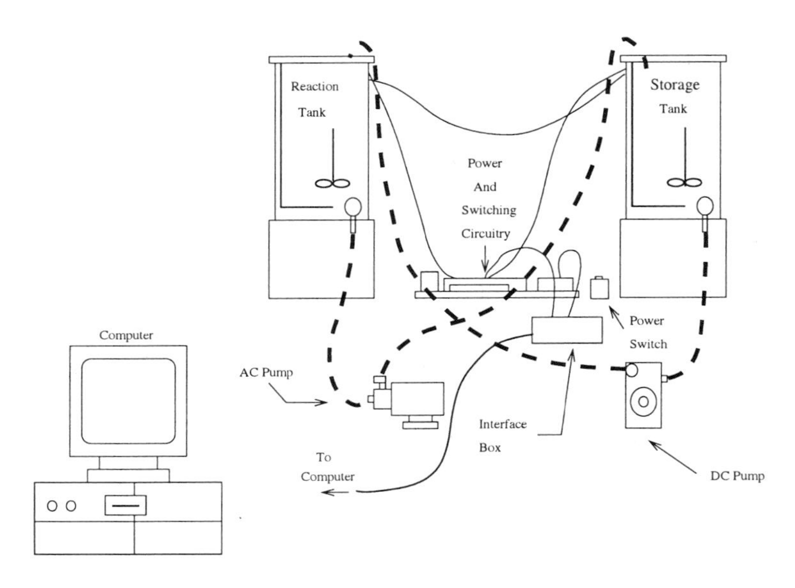

Figure 2.1: Experimental setup of the tanks.

is done by using pulse width modulation circuitry). Therefore, the DC pump is a continuous input to the plant.

The process control experiment runs on a personal computer (PC) that is called `eepc100` and can be operated both under DOS and Linux operating systems. It is possible to implement and run the control algorithms under both operating systems. For the sake of illustration, here we present only the Linux implementation. All data acquisition is obtained using a Keithley DAS20 data acquisition card. The level and temperature measuring devices are connected to the analog-to-digital converter (ADC) inputs; the mixers, the heaters, and the AC pump are controlled by the digital outputs; and the DC pump is controlled by one of the digital-to-analog converter (DAC) outputs of the DAS20 board. Note that these connections are not direct connections from DAS20 to the plant, since the outputs of DAS20 cannot provide the needed current levels. Therefore, all the connections are done through amplification circuitry, which serves electrical isolation purposes also.

Using the experimental setup shown in Figure 2.1, you can design complicated controllers which deal with both level and temperature control of both tanks simultaneously. However, our objective here is not to develop and discuss complicated control strategies for this experimental setup, as our emphasis is on using RCS for the experiment in order to provide a tutorial RCS application example. Moreover, we will not discuss in detail the plant interface and data acquisition issues. We just want to mention that in order to perform any input or output through the DAS20 data acquisition card, you need either to use the drivers provided by Keithley (which work only under DOS) or to access the DAS20 by reading from or writing to the hardware ports directly. We prefer the second option, because it also works under Linux and it is faster. In general, in order to access the ports under Linux, you need to be a root user. However, we developed a driver that gives access to the hardware ports of the DAS20 card to the nonroot programs. Moreover, we have a C++ class called `DAS20Class` that provides a library of utilities[1] to access the DAS20 card. We use the functions of this class to perform all the input and output to the plant. However, we will not show or discuss them and the related issues here.

Keeping everything above in mind, we define the control problem as follows:

> Design a hierarchical controller to control the level and the temperature in the reaction tank using the storage tank as the source for the pump that fills the reaction tank. A user on a remote computer should be able to check the status of all the modules in the system and send appropriate commands to them whenever needed.

In the rest of this chapter, we try to solve this problem using the RCS methodology and software tools. Whenever needed, we introduce the related RCS concepts and terminology and then proceed with the design. We start in the next section with a brief introduction to the RCS design methodology that is used to define the structure of the controller and how its functions are distributed.

2.2 RCS Design Methodology

Designing a practical complex control system is not generally an easy task. Moreover, difficult problems arise in maintaining such a control system. Consider a very large control system that consists of hundreds of sensors and actuators with many interdependent and interactive subprocesses. As the complexity of the system increases, so does the complexity of their comprehension, maintenance, training and operational requirements, the integrated coordination of the subsystems, and their continued operation in failure situations. In such systems the designer faces the problem of how to design a controller which will guarantee desired operation. Consider the case when one of the sensors or actuators fails, and therefore the system does not function properly. How do we find the failed component among all the components? Moreover, consider the case where a new subsystem is added to the system. Do we have to redesign the controller of the whole system from scratch? These lead to the following important questions. How do we design the control algorithm such that:

- Finding and accommodating for failures and other problems takes minimum time and effort.
- Failure of one component does not prevent the other components from functioning.

[1] These utilities were initially developed by Tony Keiser under the direction of Kevin M. Passino at OSU. See `http://eewww.eng.ohio-state.edu/nist_rcs_lib/das20.html` for information on how to download this code.

- Extensibility and upgrades of the system are easy and require minimal effort.

One possible approach is using the RCS methodology [28, 29, 33, 43, 44] that provides a framework for design of complex real-time control systems. It supports the development of a reliable system that has the ability to continue operation in the event of partial or complete failure of some of its components, or in case of loss of communication between some of its subsystems. Moreover, it supports incremental implementation and allows the extensibility of the system.

Typical RCS design starts with the study of the tasks to be performed by the system and proceeds by defining a controller structure composed of hierarchically organized control modules, which are sometimes called *control nodes.* A control module is a generic processing structure that could implement decision-making logic as well as control or estimation routines (e.g., it could implement a PID controller, a nonlinear controller, signal processing algorithms, or state table logic). The function of a control module is to perform a partial task decomposition of an input command to produce a set of simpler and easier-to-perform output commands. Task decomposition is done based on the current status of the module and its subordinates (i.e., the lower modules in the hierarchy that receive commands from this module), the present state of the internal world model database, sensory processing information, and the current command from the module's superior (i.e., the module that is higher in the hierarchy and sends commands to this module) or a human operator. Each module operates on a cycle in which it accepts commands from and reports status to its superior, sends commands to and checks the status of its subordinates, and performs sensory processing and updates (maintains) world modeling data. In each cycle operation of the module proceeds in three basic stages: *pre process, decision process*, and *post process* (refer to Figure 2.2). The start of a cycle may be triggered by a clock (periodic) or by a condition or flag in the world model (aperiodic). In Figure 2.2, the box labeled *Pre Process* holds operations that need to be performed each cycle time, independent of the current control operation and before the control operation. The box labeled *Decision Process* is where decisions take place based on some logic and appropriate action initiated. For each action that could be performed there is a function, called a *command function*, that is implemented as a state table. Note that based on the current conditions, different functions could be called each cycle time in the decision process. The box labeled *Post Process* holds operations that need to be performed each cycle time, independent of the current control action and after the control action. We discuss these functions in more detail later in this book.

There are three basic steps in a typical RCS design:

1. Task decomposition analysis.
2. Defining the controller architecture.
3. Defining the state tables for each task.

The key activities involved in these steps are: analyze tasks, define the actuators and sensors, group actuators and sensors, define control loops, define

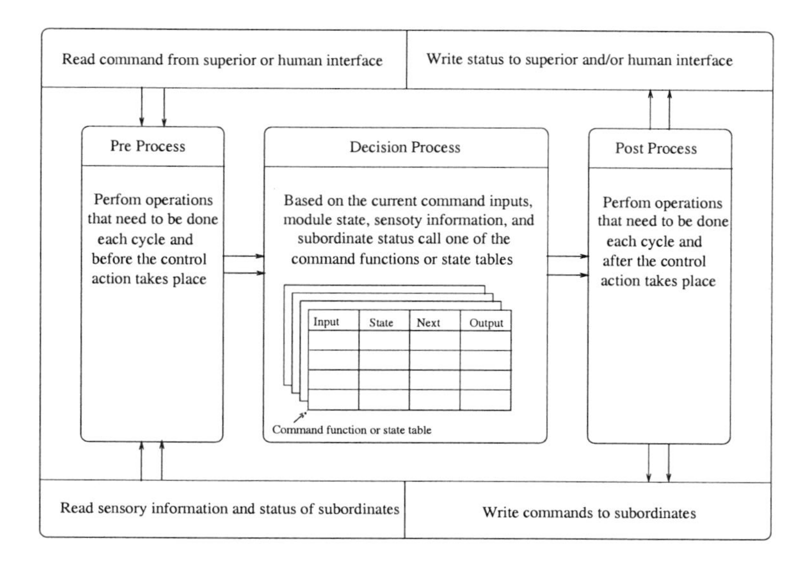

Figure 2.2: Generic control module.

state graphs (for each task), and determine inputs, status, and outputs (for each task). In very complex control systems it may not be possible to succeed in the first pass through the design of the controller. Therefore, the designer usually first designs an initial instance of the controller using these three basic steps and then iteratively evolves the design (using the same three steps again) by adding new features and functions and refining the task knowledge, module relationships, and so on. Next, we discuss briefly the three basic steps of an RCS design procedure, using the tank application as a simple illustrative example for the steps.

2.2.1 Task Decomposition Analysis

The first job of the RCS designer is to perform a *task analysis* of the physical system to be controlled. In other words, the designer must identify the tasks or operations that this system performs and which task is performed by which actuator(s) or subsystem(s). The task knowledge is represented in the form of a hierarchical task decomposition where higher-level tasks are decomposed into lower-level subtasks (i.e., where the execution a sequence of subtasks results in achieving the high-level task). For instance, consider the task "Produce chemical A" shown in Figure 2.3. Assume that this task requires that chemicals B and C are mixed with rates n and m at temperature T and pressure P to form chemical A. Then one possible decomposition of this task could be as shown in the figure. After that, each subtask of this task could be decomposed further to even simpler subtasks. This procedure is sequentially repeated starting from the top, until the bottom actuator outputs are reached. Then, you need to analyze

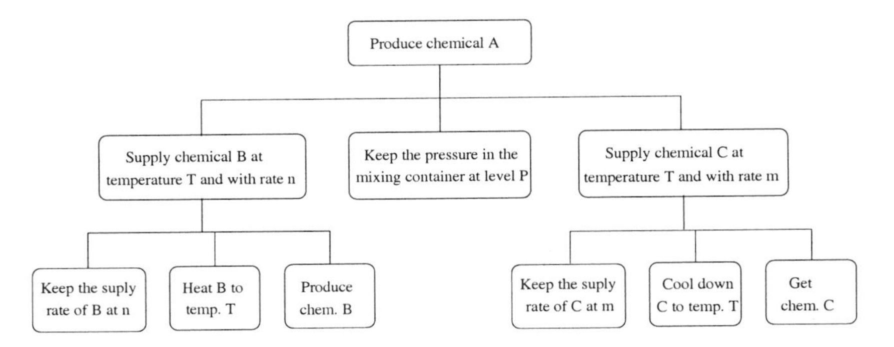

Figure 2.3: Example of task decomposition.

the possible task sequence scenarios of the desired operations and try to choose the best one. To this end, the designer may need to refine the scenarios to the level of detailed interactions between actuators and sensors. This also aids in refining the task parameters and knowledge (i.e., the data needed to represent the tasks) and it breaks the problem down into simpler problems since every task can be considered as a separate small-scale control problem, which needs certain conditions to be met in order to be initiated, and produces some outputs or goals when completed. The goals of one task can be initiation conditions for other tasks.

To illustrate the procedure, we will perform a task decomposition analysis for the process control experiment. To start, review the experimental setup and the control problem described in Section 2.1. For the reaction tank we have two sensors and four actuators. The sensors are for measuring the level and temperature of the reaction tank and the actuators are the heater, two pumps for filling and emptying the tank, and the stirring mechanism in the tank. The stirrer is used in order to counteract the disturbance due to the high-pressure pumping (this allows for a more accurate reading in our level sensor) and also to mix the liquid so that there is no temperature difference in different sections of the tank. As discussed before, we know that all the actuators could be either on or off. If pumps are on, they pump water either in or out of the tank. In other words, they either increase or decrease the level of the liquid. If the heater is on, it increases the temperature of the liquid, and if the stirrer is on, it counteracts the turbulence (to some extent) and homogenizes the temperature within the tank. In the problem stated, the role of the storage tank is only to provide a container for storage of the liquid for the pumps in order to add or remove water from the reaction tank.

The objective is to regulate the level and temperature in the reaction tank. Therefore, the highest-level task for this system would be "Regulate the level and temperature of the liquid in the reaction tank at desired level and desired temperature." This task could be broken down to two subtasks named "Regulate the level of the liquid in the reaction tank at the desired level" and "Regulate the temperature of the liquid in the reaction tank at the desired temperature." The task "Regulate the level of the liquid in the reaction tank at the desired level" can be decomposed into "Measure the liquid level," "Increase the liquid level to achieve the desired level," and "Decrease the liquid level to achieve the desired level." We can continue similarly for the temperature and obtain the decomposition shown in Figure 2.4. Note that some of these tasks

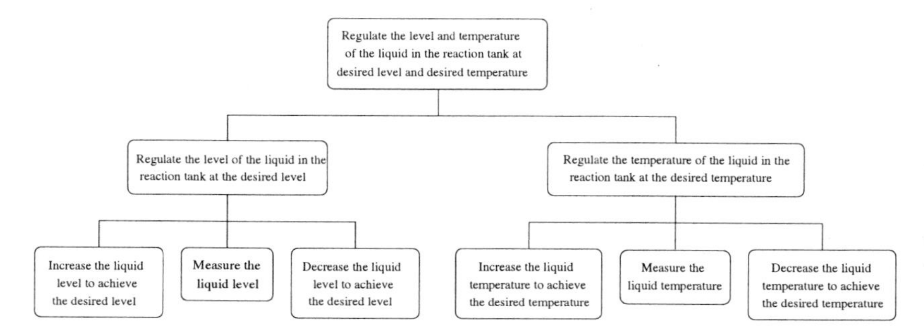

Figure 2.4: Task decomposition for the process control experiment.

are concurrent and should be in mutual exclusion for proper operation of the plant. For example, we should not have the tasks "Increase the liquid level to achieve the desired level" and "Decrease the liquid level to achieve the desired level" initiated simultaneously. In fact, the initiation of these tasks depends on the data obtained from the task "Measure the liquid level," and the task "Increase the liquid level to achieve the desired level" is initiated if the current level in the tank is less than the desired level, whereas the task "Decrease the liquid level to achieve the desired level" is initiated if the current level in the tank is greater than the desired level.

Note that we reached the lowest level of the task decomposition hierarchy, since each task on the lowest level corresponds to a simple job of a particular sensor or actuator. For example, the task "Measure the liquid level" will be performed directly by the level sensor, whereas the task "Increase the liquid level to achieve the desired level" will be performed by the DC pump. Note also that for the task "Decrease the liquid temperature to achieve the desired temperature" we do not have any actuator in the system (i.e., we do not have cooling system in the setup). It is possible to decrease the temperature by pumping cold water from the other tank but we will not deal with these issues here. Therefore, in cases when there is a need to decrease the temperature, we will rely on the ambient conditions (i.e., the heater is simple turned off).

2.2.2 Defining the Controller Architecture

For defining the hierarchical architecture, the first and most important thing to consider is the layout of the physical subsystems and all the actuators and sensors of the system to be controlled. Each subsystem will have its own sensors and actuators. Then, based on the physical layout of the subsystems, the connections between them, the information flow, and the task decomposition analysis performed in the previous step, we define the controller architecture. This typically starts with assigning a control module to each actuator and sensor on the bottom of the hierarchy and continues by defining the modules in the upper levels based on the results obtained from the higher-level task decomposition analysis and subsystem delineation. You may consider this step as mapping the task decomposition hierarchy obtained in the previous step into a organizational hierarchy of implementable control modules. At this stage we already know the areas of activity and responsibility of each module. We know which commands each module accepts and from which module, which commands it can send and to which modules, which commands are decomposed to which subcommands and under which conditions, which sensory information or status data from the subordinates the modules need for successful operation, which data the command and status messages contain, and so on. Defining the controller in a hierarchical way leads to an information and control structure which is easy to manage even in fairly complex systems. Note that the definition of all the command, status, and other information needed for successful operation of the module also defines the data representation needed by the world model kept inside of the system.

At this step we may also define the operator interactions with the system (if there is a need for any). In other words, we may define which modules we can interact with, which commands we can send to them, and which status information we can observe or monitor in order to carry out our decisions (which may be crucial for proper operation of the plant). In general, the operator interacts with the modules at the higher-level of the hierarchy and performs high level decision making. However, in some cases you may interact with the modules at the lower levels also.

Consider, once again, the process control experiment. Following the design guidelines, we can begin the design of the controller by assigning a control module to each actuator and sensor at the bottom of the hierarchy (see Figure 2.5). Then, note that we have two independent operations of temperature and level regulation that are performed by two different subsystems. Therefore, we can assign a module as a coordinator to each of these subsystems at one level higher in the hierarchy (i.e., the level regulator and the temperature regulator modules in Figure 2.5). Finally, we assign a supervisor on the top of the hierarchy to coordinate the actions of all the modules in the system and hence obtain the hierarchy shown in Figure 2.5. This is a fairly simple hierarchy; however, we would like to simplify it even more. (The reason for this is the fact that we want to reuse the code that we had developed before for this system.) Our simplification consists simply of removing all the modules in the lowest level of the hierarchy and assigning their job to their coordinator on the level above it. In other words, the level regulator module, or level module, will have control of the two pumps, direct access to the readings of the level sensor, and will perform the tasks of increasing and decreasing (i.e., regulating) the level to (at) a specified set point. Meanwhile, the temperature regulator module (or, with its new name, the heater module) will have control of the heater and direct access to the readings of the temperature sensor and will perform the task of

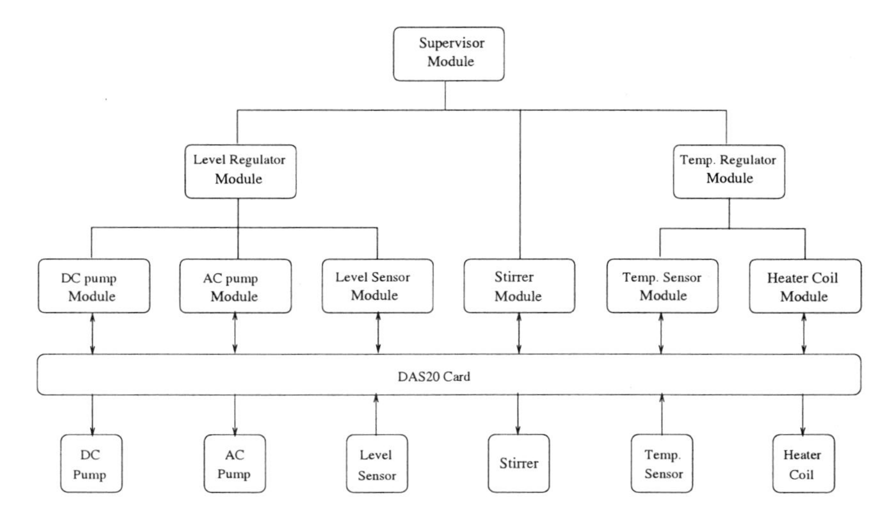

Figure 2.5: Initial controller hierarchy for the tanks.

increasing and decreasing (i.e., regulating) the temperature to (at) a specified set point. Moreover, we let the stirrer always be on as long as the program is running and hence remove its control module from the hierarchy. The justification for this is that with this number of sensors in the system it is difficult to develop an algorithm which will successfully report that the temperature of the liquid is homogeneous enough so that if the filling pumps are not running, we stop the stirrer. In other words, we do not have any logic to start or stop the stirrer. Moreover, we do not have a sensor to tell us whether the stirrer is on successfully or if it stops because of a hardware problem. In other words, the stirrer does not have status to report. Therefore, it is a good idea to keep it on as long as the program is running and there is no need to develop a control algorithm for it. Furthermore, this also illustrates the flexibility of RCS design (i.e., the freedom of the RCS designer in selecting the hierarchy of the controller). In other words, we can choose any structure of the controller that will meet the control objectives. After these simplifications, the hierarchy of the controller becomes as shown in Figure 2.6, where we also show the operator interface to the modules with dashed lines. To be consistent with the C++ code that we discuss later, sometimes we refer to these modules as `SUPERV_MODULE`, `LEVEL_MODULE`, and `HEATER_MODULE`.

Each of the three modules in Figure 2.6 is of the form of the control module in Figure 2.2. To define more fully how the modules interact with the tank system, the operator, and each other, we need to define more fully the communications that can occur (e.g., commands and status). We define a command called `LEVEL_SET_REF`, which corresponds to a request of the task "Regulate the level of the liquid at the desired level" in Figure 2.4. This command will be sent by the supervisor (or human operator) and accepted by the level module. Note that this command should carry information about the numeric value of

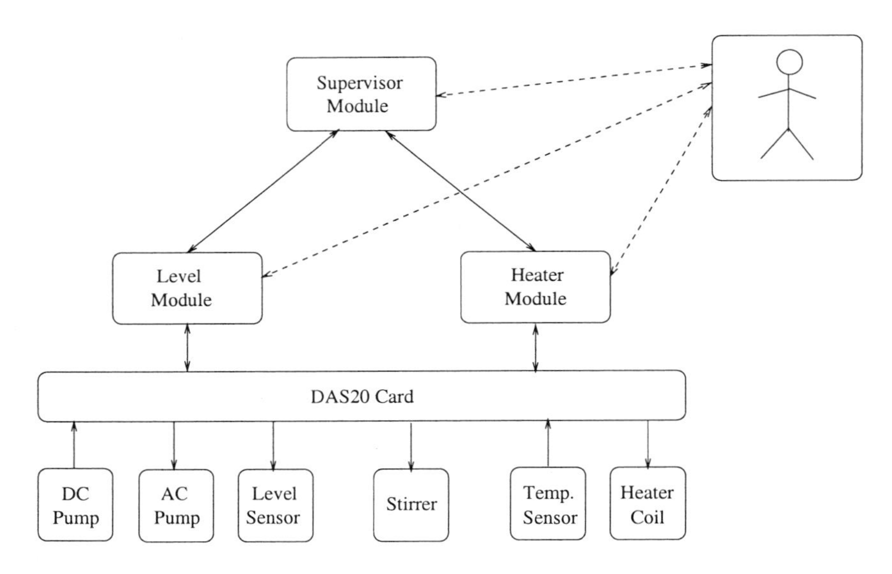

Figure 2.6: Simplified controller hierarchy for the tanks.

the *reference level* to be set (i.e., the *desired level*). The level module must then use the measured liquid level to increase or decrease the liquid level appropriately to meet the demands of LEVEL_SET_REF command. In other words, the corresponding function of the LEVEL_SET_REF command should contain an appropriate control algorithm to turn the pumps on and off for regulating the level of the liquid at the *desired level.*

Similarly, we define the command HEATER_SET_REF, which corresponds to a request of the task "Regulate the temperature of the liquid at the desired temperature" in Figure 2.4. This command also will be sent by the supervisor (or human operator) and will be accepted by the heater module and will carry information about the *desired temperature.*

In order for the supervisor to be able to make correct decisions, it should know the current temperature and level. Therefore, we define the status message of the level module, LEVEL_STATUS, with a data field for *current level.* Similarly, the status message of the heater module, HEATER_STATUS, has a data member for the *current temperature.*

Corresponding to the the task "Regulate the level and temperature of the liquid at the desired level and desired temperature" in Figure 2.4, we define the command SUPERV_SET_REF for the supervisor. This command will be accepted by the supervisor and can be sent by a human operator and will have two data fields for both *desired level* and *desired temperature.* When the supervisor receives this command it sends the commands LEVEL_SET_REF and HEATER_SET_REF to the level and heater modules, respectively. In this way a human operator will have access to the plant and will be able to change or set reference points through the system. Moreover, the supervisor status message,

SUPERV_STATUS, will contain data fields for both the *current level* and *current temperature* in order for the operator to be able to view them. We can limit the interactions of the operator to the plant to his or her interactions to the supervisor; however, as a design objective we need the operator to be able to have access to the other two modules, too. Moreover, in some situations it may be necessary for the operator to be able to interact with some modules in the lower levels of the hierarchy. Therefore, we also set channels between the operator and the other two modules so that we can send commands (those that each module accepts) directly and check the status of the level and heater modules also.

Continuing with the command definition, we define the INIT and HALT commands for each of the three modules in the system. The task of the INIT command is to initialize the system in order to start from a known state. The objective of the HALT command is to stop all related actuators safely and halt system operation. We summarize all the command and status messages in Table 2.1.

2.2.3 Defining State Tables

After task analysis and controller architecture definition, the designer knows which operations can be performed by the system, which subsystem or module will perform which task, which scenarios lead to a desired performance, and what information the modules in the hierarchy need to share. To continue with the design, the designer breaks down each task to subtasks or state tables using *state table analysis.* State table analysis can be done by breaking down the task to a sequence of operations in time and space which represent different "states" of the system. For each of these states, the sensory information, the current status of the module, the operator requests, the status of the module's subordinates, and so on, are determined. Generally, state tables are defined for the modules at the bottom first and then they are constructed successively for higher levels.

Consider once again the process control experiment. Let us identify the different states of the commands in the system (i.e., we will perform a state table analysis). We will perform this analysis for one command, and the analysis for the other commands is similar. Consider the command LEVEL_SET_REF. When this command is received by the level module, it checks whether the *current level* is greater or less than the *desired level* specified by the command, then it starts either filling or emptying the tank. Therefore, the input to this task is the desired level and the output of the task is the current level in the tank. For safety purposes, we also define minimum and maximum *safety levels* for the tank (we do not want the tank to empty totally, or overflow). Therefore, we identify four different states for this command: *fill (s1)*, which occurs if the level is too low; *empty (s2)*, which occurs if the level is too high; *idle (s3)*, which occurs if the level is just about right; and *error (s4)*, which occurs if the level exceeds the safety levels. The state transition diagram for these states is shown in Figure 2.7. This leads to the state table shown in Table 2.2, where we show

Table 2.1: Message types for the process control experiment ("cmd" indicates a command, "sts" indicates a status).

Message Name	Type	Data Fields	Description
SUPERV_INIT	cmd		Initializes process control experiment (both level and temperature controllers).
SUPERV_HALT	cmd		Stops level and temperature controllers and halts overall system operation.
SUPERV_SET_REF	cmd	desired_level desired_temperature	Sets reference temperature and level values in lower modules.
SUPERV_STATUS	sts	current_level current_temperature	Provides supervisor status information.
LEVEL_INIT	cmd		Initializes level control algorithm.
LEVEL_HALT	cmd		Stops level control algorithm.
LEVEL_SET_REF	cmd	desired_level	Sets level reference value used in control algorithm.
LEVEL_STATUS	sts	current_level	Passes level status information to supervisor and interface.
HEATER_INIT	cmd		Initializes temperature control algorithm.
HEATER_HALT	cmd		Stops temperature control algorithm.
HEATER_SET_REF	cmd	desired_temperature	Sets temperature reference value used in control algorithm.
HEATER_STATUS	sts	current_temperature	Passes temperature status information to supervisor and interface.

the conditions for the state to occur, the action taken in this state, the reported status for each state, and the next possible state, together with the conditions for transition to that state.

Following a similar approach, we can define the state tables for the other commands or tasks to be performed by each of the modules in the system. Note from the table that during execution of the command LEVEL_SET_REF, the system is at only one state at any given time instant. In other words, there is no overlap between states. This should be true for all the commands in the system. This allows the remote user to monitor the current state through the RCS diagnostics tool (discussed later). Moreover, in each cycle time only the code for that state is executed, hence reducing the computation time. For this reason, even in fairly complex systems the computation needed by a given module in

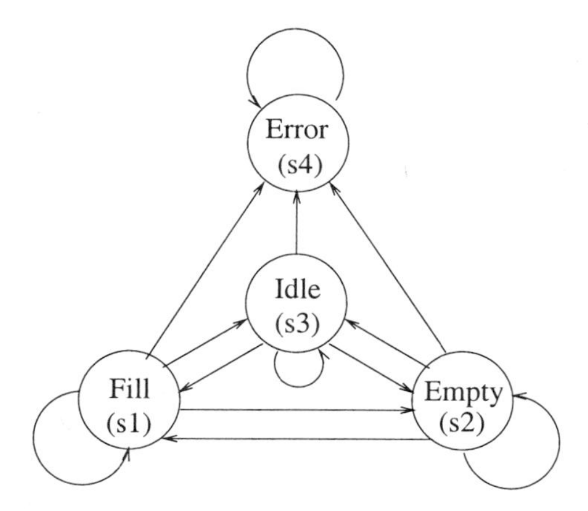

Figure 2.7: State transition diagram for the LEVEL_SET_REF task (command).

each cycle time is relatively small. Moreover, note that the comprehension, development, and modifications of the control algorithm for each of the states is fairly simple. At the lowest-level modules, such as the level and heater modules, they could be simple off/on decisions or PID (proportional-integral-derivative) control loops.

2.2.4 Implementing the RCS Controller

We have completed the design of the RCS controller for the process control experiment. Since the design is complete, all the task knowledge, sensory information, command and status message vocabulary, interactions between the modules and the information to be shared, the scenarios of successful operation, and the operator interactions with the system are known. Within this framework, all the modules in the system can work independently while being connected via a communication system. Now, assume that for some reason the level sensor or perhaps one of the pumps failed, so that the level module is unable to perform its job. The heater module can still run and increase or decrease the temperature of the liquid to the desired temperature. Moreover, a (remote) operator can immediately determine, through the RCS diagnostics, that the level module has a problem, and identify the failed component. If on the other hand, for some reason there is a loss of communication between the modules, each module can have an "emergency routine" and continue its operation through this routine and implement a predefined control algorithm, invoke an alarm to call attention to the problem, or safely stop and shut down the system. Moreover, consider the case in which new subsystems (e.g., new tank and pump) are added to the system. We can very easily reuse the algorithms that we developed for the first tank in the new tank. Moreover, we do not have to redesign the whole controller, but only design the modules for the new subsystem and define the interactions of these modules with the other modules.

Table 2.2: State table for the LEVEL_SET_REF command (cl = current level, dl = desired level, min = minimum safety level, and max = maximum safety level).

State	Conditions	Action	Status	Next State
fill (s1)	min $<$ cl $<$ max and cl $<$ dl	DC pump on AC pump off	EXEC	s1 (cl $<$ dl) s2 (cl $>$ dl) s3 (cl $\cong$ dl) s4 (cl $<$ min or cl $>$ max)
empty (s2)	min $<$ cl $<$ max and cl $>$ dl	AC pump on DC pump off	EXEC	s2 (cl $>$ dl) s1 (cl $<$ dl) s3 (cl $\cong$ dl) s4 (cl $<$ min or cl $>$ max)
idle (s3)	min $<$ cl $<$ max and cl $\cong$ dl	DC pump off AC pump off	DONE	s3 (min $<$ cl $<$ max) s1 (cl $<$ dl) s2 (cl $>$ dl) s4 (cl $<$ min or cl $>$ max)
error (s4)	cl $<$ min, or cl $>$ max, or any other error	DC pump off AC pump off Report error	ERROR	s4

Having finished the design, we face the challenge of converting this design to a computer code or, in other words, to implement it. The NIST RCS software library has been developed for this purpose. Using the RCS design tool, you can easily lay out the modules in the current design and generate all the application-independent code and then fill in the application-dependent control or estimation algorithms and data variables. In the rest of this chapter, we deal with implementation of the RCS controller using the NIST RCS library. In the next section, we describe some of the essential components of the RCS library and establish the related terminology and concepts. After that, we describe the step-by-step use of the RCS design tool to generate the code for the process control problem, and then we explain some of the code to complete the development of the tank control problem.

2.3 RCS Software Library Essentials

In this section we describe some of the essential features of the RCS software library and establish the necessary terminology.

The RCS library was developed based on the RCS methodology and the Reference Model Architecture (discussed in Chapter 3). It can be used for implementing hierarchical controllers running simultaneously on the same or different computers. These computers may be running using different platforms and therefore may have different internal representations of data. Usually, the modules in the controller need to communicate with each other. The communications between these processes occur by passing information via common shared memory buffers using the CMS (Communication Management System) and NML (Neutral Message Language) routines. CMS and NML are also responsible for encoding and decoding data so that it can be interpreted by otherwise incompatible platforms. The terms *command buffer* and *status buffer* refer to memory buffers that are devoted to sending commands and the current status of the processes, respectively. In a general RCS hierarchical structure, controllers which are higher in the hierarchy in general should have access to the command and status buffers of the lower-level processes. That way, commands are passed to the lower subsystems from the top of the hierarchy, and the status of those lower systems can be monitored by the higher-level structures.

2.3.1 Overview of CMS and NML

The CMS is the low-level communications system in RCS. It consists of a number of operating system calls that allow it to gain information about host computers so that processes running on different platforms can communicate with each other. The RCS programmer will not have to deal directly with the lower-level utilities of CMS, since NML provides a convenient high-level interface to the utilities. However, it is important to understand what is going on within CMS in general, since this will aid in finding solutions to any problems that might occur in implementation.

The CMS provides access to a fixed-size memory buffer of general data to multiple reader or writer processes on the same processor, across a backplane, or over a network. This memory buffer simply acts as a shared portion of space, through which multiple processes "talk" with each other by writing messages into the buffer for other processes to read. The size and location of this memory buffer is user-configurable using an NML configuration file discussed in later chapters. In general, these memory buffers can exist on any computer connected to the RCS chain. However, remote processes (processes that are not located on the same computer as the memory buffer) write to and read from the buffer using a server.

CMS is the communications "workhorse" of RCS. Not only does it provide access to the buffers, it also keeps track of what is going on with the buffers, such as whether or not the buffer has been written to or read from, and if the data currently located within the buffer is new or old. This is accomplished through a set of CMS C++ classes. A header is added to each buffer by the CMS, which provides the information above as well as if the buffer is new to a particular process and the size of the last write to the buffer. As stated before, the user will not access many of the CMS functions and status variables directly, but will access this information via NML functions. In fact, most of the NML classes are simply derived from the CMS base classes, and therefore, predefined

NML functions from those classes give the programmer access to the status of the buffers.

CMS also provides a very general communications interface. Regardless of the communication methods required by a particular process, the interface to CMS is uniform. In fact, the developer can even change communication methods, buffer sizes, and other communications parameters from the NML configuration file (discussed in detail in Chapter 7), without having to recompile or relink their code. Methods are provided to encode all the basic C data types in a machine-independent or neutral format, and to return them to the native format so that communications can occur across different operating systems. Since all CMS methods understand the neutral format, data written to a buffer from an RCS process, say, on a UNIX platform, can be read by any other RCS process regardless of platform type (Windows, DOS, etc.) provided the process has access to that buffer. The NML, then, provides a mechanism for handling multiple types of messages in the same buffer as well as simplifying the interface for encoding and decoding buffers in neutral format and the configuration mechanism.

NML provides a convenient interface to CMS to improve interaction and keep the developer from having to deal with platform-specific issues. It also contains generic operating system-independent read and write functions to access memory buffers. These generic calls actually point to specific operating system calls within the CMS system. Which calls are actually used depends on the platform the process is running on. Perhaps most important to the developer, NML contains a generic controller module, called `NML_MODULE`, that contains functions that allow for preprocessing and postprocessing of data, interaction to buffers, access to state-table functions that ease the design of multi-state systems, and so on. The `NML_MODULE` is the key class structure in RCS and often through this document we use the term NML module, RCS module, RCS control node, or the name `NML_MODULE` to mean a controller (actually, subsystem) written in functional RCS form. Note that the term *controller* is used quite often in reference to RCS modules. Keep in mind that RCS can also implement estimation and prediction routines as well as virtually any decision-making process.

The NML functions and classes allow the user to develop a rather wide range of system structures. But it also provides the developer with the ability to design the contents of the messages that are written to the buffers by each process. C++ classes exist to develop both command messages (which pass commands to lower subsystems) and status messages (which return the status of the system to higher subsystems or modules). This allows the developer to determine which modules are able to perform which commands, as well as to determine the necessary information that should be passed along to other RCS processes.

2.3.2 NML Application Structure and Terminology

Figure 2.8 illustrates the structure of a typical RCS application using NML. The application is distributed across three computers. Here, we see that the memory buffers used for message communications are established on computer 1. Processes 1, 2, and 3 are able to write directly into the shared memory buffers they use because they are located in the same computer or backplane. It is for this reason that they are labeled LOCAL. Processes 4, 5, and 6 can only access the buffers through an NML server and are therefore labeled REMOTE. Shared memory buffers can exist on multiple machines as well. If this is the case, then the description can be more complicated; the processes would need to be described as local or remote with respect to a particular buffer.

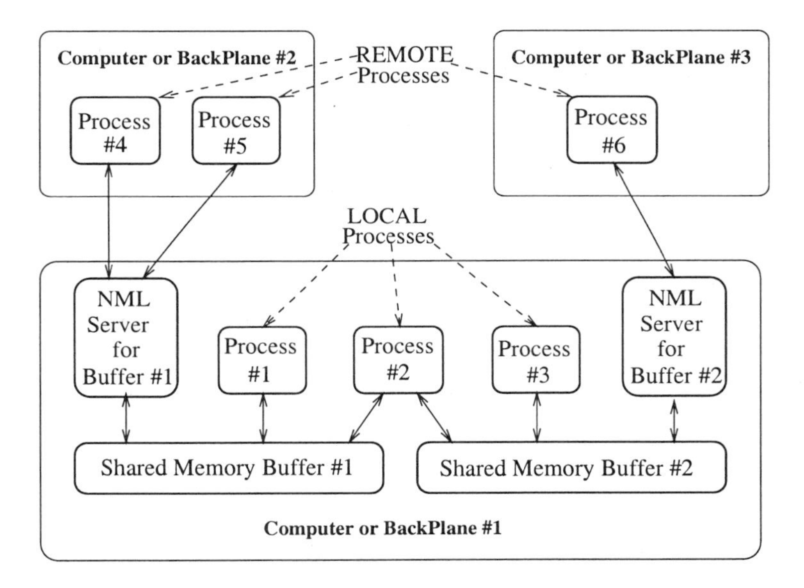

Figure 2.8: Example of structure of typical RCS application using NML (from [46]).

It is necessary to run NML servers for each buffer that will be accessed by remote processes. These servers read and write to the buffer on behalf of remote processes in the same way that local processes do. NML uses configuration files to store information about which processes communicate with which buffers and how. Most of the options available to NML programmers are chosen by specifying them in the configuration file. Note that CMS and NML use the same configuration file.

Reading and writing to memory buffers are message-based operations in RCS. This allows a read operation to retrieve an entire message at once (at least from the viewpoint of the developer). As such, each successful read operation retrieves the data sent in exactly one write operation, and unless queuing is enabled, each write operation moves one message into the buffer, replacing any previous message. More than one type of message can be sent to the same buffer, so a unique identifier is always contained in the message. After a read operation, the process must use this identifier to determine the type of message before using any of the data in the message. Each type of message implies a particular data structure. Most messages are user-defined.

Messages are called *encoded* if they have been translated into a machine-independent or neutral format such as the *eXternal Data Representation (XDR)*. Buffers are called encoded if the messages in them are to be encoded and this is established in the configuration file. NML servers can encode and decode messages on behalf of remote processes. An NML vocabulary defines the set of messages that may be used in an application and provides the necessary functions for encoding and decoding messages.

Because NML is configurable, programmers can choose between protocols with higher performance but which may be more restrictive or require more expensive hardware, and those that are less restrictive or require less expensive, more widely available hardware. By making a buffer local to a process you can improve the performance of that process. By moving processes you may be able to reduce the load on one CPU (central processing unit) or increase the number of processes able to use the faster local protocol. Using servers to provide remote access to buffers frees local processes from being slowed down by the communications with remote processes.

Here is a summary of design suggestions that the developer should take into account while designing an NML application:

1. Avoid overloading any CPU by assigning too many processes to it, or building a single process which must do too much work.

2. Place buffers so that they may be accessed locally by the most time-critical process(es).

3. Use the LOCAL protocol whenever possible.

4. Use neutrally encoded buffers only when necessary (i.e., backplane communications between different types of processors).

Now consider the process control experiment. In the preceding section we designed an RCS controller for the process control experiment with structure shown in Figure 2.6. In this section, we show how this structure is implemented using the RCS library. During design for the process control problem we decided that a remote user will have access to all the modules. Implementing of this remote user interface provides us an opportunity to demonstrate to the reader the over-network communication tools of the RCS library.

From Section 2.1 we know that this experiment can be run both as a DOS or Linux executable on a PC (`eepc100`). In the lab we have a Windows NT workstation (`eepc99`) and other Linux machines (`eepc73, eepc97, eepc98`) that we can use for diagnostics purposes. We choose the computer running under Windows NT for diagnostics since this implements communications between computers with different operating systems and serves as an illustration of the platform independence of the RCS library.

If we run the application under DOS, we will need to put all the NML buffers on the diagnostics computer (i.e., the NT workstation or one of the

Linux machines). This is because we need to run NML servers for them and DOS does not allow multitasking. (On `eepc100` there is already the controller process running; therefore, we cannot run the NML servers there.) For a Linux application, on the other hand, it is possible to set up the buffers in both of the computers. We prefer to put them on the computer running the controller, since this makes it easy for the process to write to the buffers, which speeds up the process because it will not incur the associated network communication delays.

Figure 2.9, where the Linux PC implements the tank controller and the Windows NT PC is the user interface computer, illustrates the NML structure for this choice. In the three modules of the application each process is allocated two

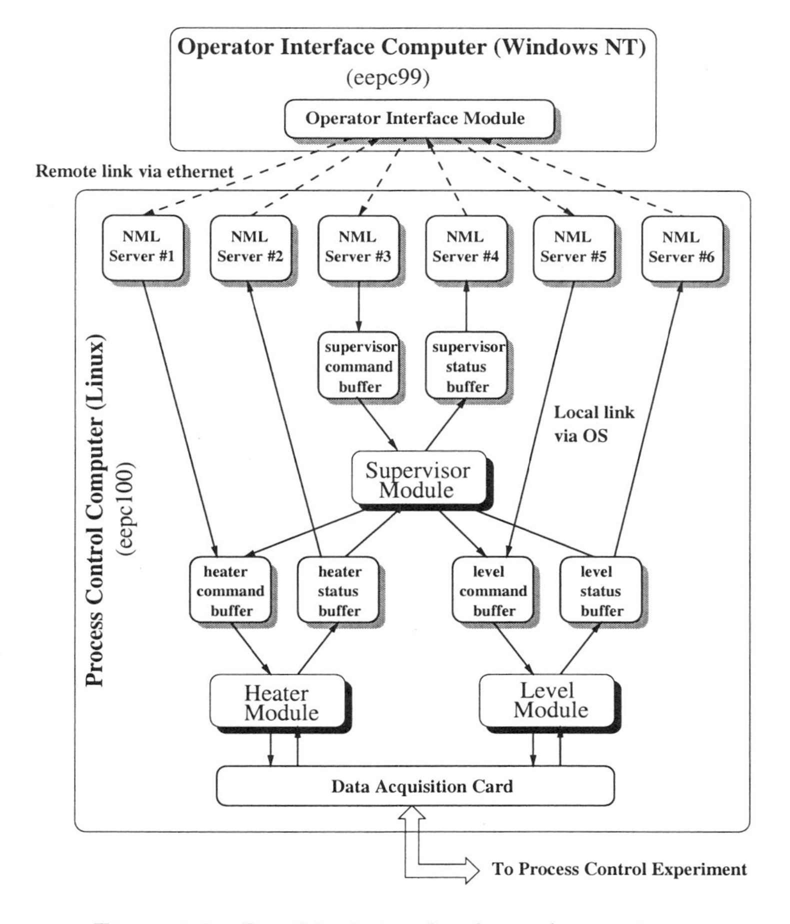

Figure 2.9: Possible design for the tank experiment.

memory buffers—one for status information and one through which commands are passed—and the modules' access to each buffer depends on what information needs to be shared (refer to Figure 2.9). It is obvious that each process (module) needs to access its own buffers. Since the supervisor will be sending commands to the subordinate level and heater modules, it needs access to those modules' command buffers. Status information of the subordinates is passed to the supervisor as well, so the RCS architecture gives the parent module access to those buffers as well.

We chose the structure in Figure 2.9 to make it possible for the controller modules to directly read from or write to the shared memory buffers, so that reading command or writing the status takes a minimum amount of time. Another possibility could be to define all the buffers on the computer for the user interface and run the NML servers on that computer. (This would be necessary under the DOS operating system because of the reasons mentioned before.) The problem with this design would be that it would slow down the controllers, because it will require more time to write and read the shared memory buffers.

2.3.3 Structure of a Single Process Using NML

The RCS library classes which provide the programming interface for CMS and NML are `CMS`, `CMS_HEADER`, `CMS_SERVER`, `CMS_USER`, `NML`, `NMLmsg`, `NML_SERVER`, `RCS_CMD_CHANNEL`, `RCS_STAT_CHANNEL`, `RCS_CMD_MSG`, and `RCS_STAT_MSG`. Moreover, the RCS library has a class, called `NML_MODULE`, that provides a generic controller module class. It contains in itself also all the tools for establishing the NML communications with other modules by providing functions to create NML communication channels and send NML messages.

Many of the classes above are not used explicitly in simple RCS implementations (or even fairly complex ones). The most important classes that the user will encounter in almost any application are:

- `NML_MODULE`,
- `NML`,
- `NMLmsg`,
- `RCS_CMD_CHANNEL`,
- `RCS_STAT_CHANNEL`,
- `RCS_CMD_MSG`, and
- `RCS_STAT_MSG`.

In order to use these classes you need to include the header files for these class definitions. Including `rcs.hh` in an RCS application program results in the inclusion of all of the necessary header files. In Part II we describe the NML classes as well as other classes of the RCS library in detail.

Figure 2.10 shows the structure of a single concurrent process module using NML. Memory buffer 1 appears to be local to the application since it is writing directly to this buffer, and memory buffer 2 is remote to this module because it accesses it via an NML server. The figure depicts what has been discussed earlier, specifically that the developer is interested primarily in the NML portions for applications development. The circled items—NML library interface, CMS update functions, and CMS communications functions—are all modules from the RCS library and need not be written by the programmer. Only the application-specific functions are user-written (or some of them could be generated by the RCS design tool, as we will see soon).

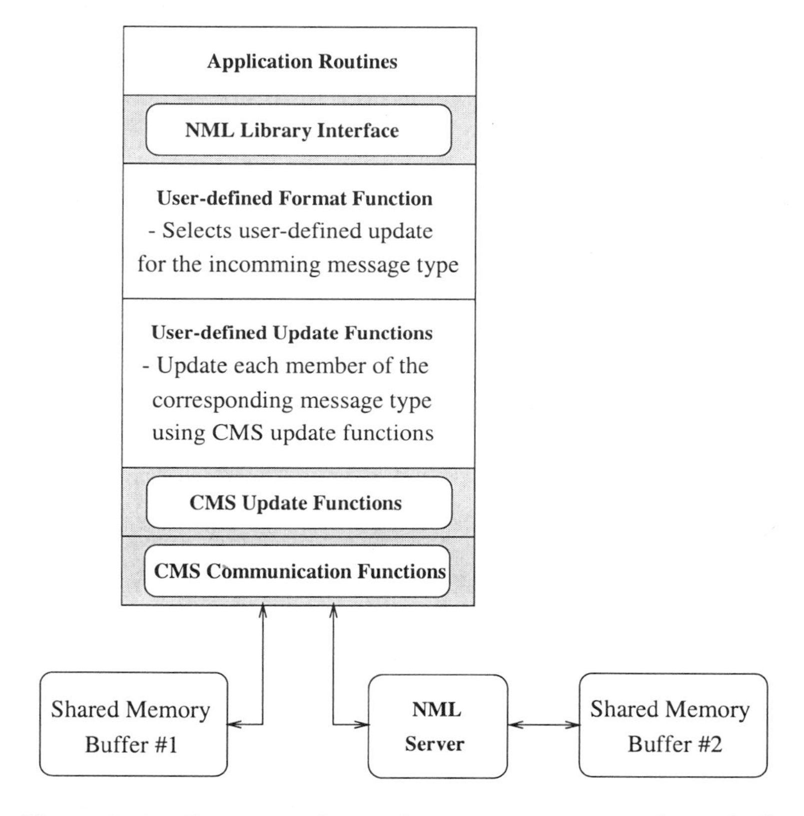

Figure 2.10: Structure of a single concurrent process (from [46]).

The applications routines initialize and use objects from classes `NML` and `NMLmsg` which depend on some user-defined functions. The content of these classes and how to write them are discussed in Chapter 5. For now, it is only necessary to understand that NML classes are for establishing communication channels (i.e., access to particular buffers) and NML message classes are for passing information to other processes through the NML communication channels (i.e., the data to be written to these buffers). RCS programmers are responsible for providing a *format function* together with *update functions* for each type of message, which will be passed to the memory buffer (the two unshaded boxes in the middle of Figure 2.10). The format function selects the appropriate update function routine for each message type that exists. The update function is the primary, and in most cases the only, direct interaction an RCS programmer needs to make to CMS. The primary purpose of this function is to update the values of the data members of the messages in the internal CMS buffers. Recall that reading and writing to memory buffers actually involves low-level operating system calls that are handled by CMS. The actual read and write functions available in the NML classes simply point to the appropriate low-level CMS functions. Since it is the CMS which actually passes the messages between buffers, a function must be available that updates the values as seen by the CMS so that correct values of the data variables are passed on to the shared memory buffer. The update function accomplishes this task. Furthermore, CMS already provides update functions for the basic C data types (floats, integers, characters, etc.); the update function for each message, then, can be built simply by updating each member (variable) of the message individually using the CMS update routines for the basic C data types. These basic update routines write to and read from internal CMS buffers which are themselves read or written to memory buffers that are available to other concurrent processes using the CMS communications routines.

2.4 Generating Code via the RCS Design Tool

In this section we show how simple it is to generate most of the application code for the process control problem, or any other application, using the RCS design tool. The RCS design tool and other code generation tools of the RCS library are described in more detail in Chapter 10.

The RCS Design tool is a Java-based graphical program which can be used to easily define the modules in a controller hierarchy, define the command messages for the modules, group the modules to particular execution loops with needed cycle times, assign NML servers to particular buffers, and generate most of the application code automatically, including the scripts for compiling and running the application.

The design tool can be viewed using any Java-compatible Web browser, and can be run as an applet or as a stand-alone application. To run it as a stand alone application (on UNIX based systems), type

```
java -cp $CLASSPATH rcsdesign.rcsDesign
```

on the command line. Above, the environmental variable `CLASSPATH` should be set to the path for the RCS Java library and the Java Development Kit library. For example, in Linux the RCS Java library is usually installed under the directory `/usr/local/rcslib/plat/java/lib`, whereas the Java Development Kit library is under `/usr/local/jdk/lib`. To set an environmental variable you can use the `setenv` (or `set`) command under UNIX-based systems.

2.4.1 Specifying the Hierarchy and Generating the Code

When you start the design tool the screen will look like the one shown in Figure 2.11. Then you can use the "Open" button (in the upper right-hand corner

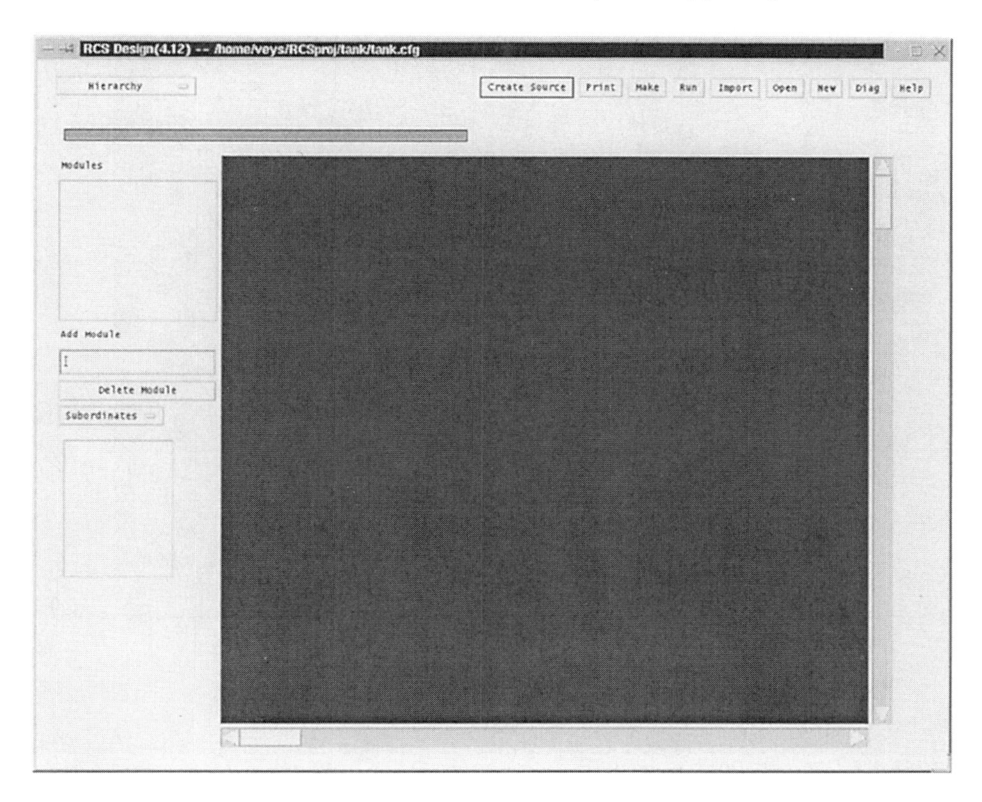

Figure 2.11: Initial screen of the RCS design tool.

of the screen) to open an existing application or the "New" button to start a new design. We press the "New" button and provide the name `tank` for the name of the application and the path `/home/veys/RCSproj/tank` for both user and application directories, when asked. After that we can start building the hierarchy of our controller. This is done by typing the name of the modules in the box labeled "Add Module" on the left side of the screen. Each new module is added as a subordinate of the current module selected from the modules list on the left of the screen. You can rearrange the hierarchy by choosing a module from the modules list and then choosing its subordinates from the subordinates list. The hierarchy of the tank controller established in Figure 2.6 is shown in Figure 2.12. Note from the figure that on the modules list the supervisor is highlighted and on the subordinates list the level and heater modules are highlighted, meaning that the level and heater module are subordinates of the supervisor module. This is also reflected visually in the hierarchy drawn in the

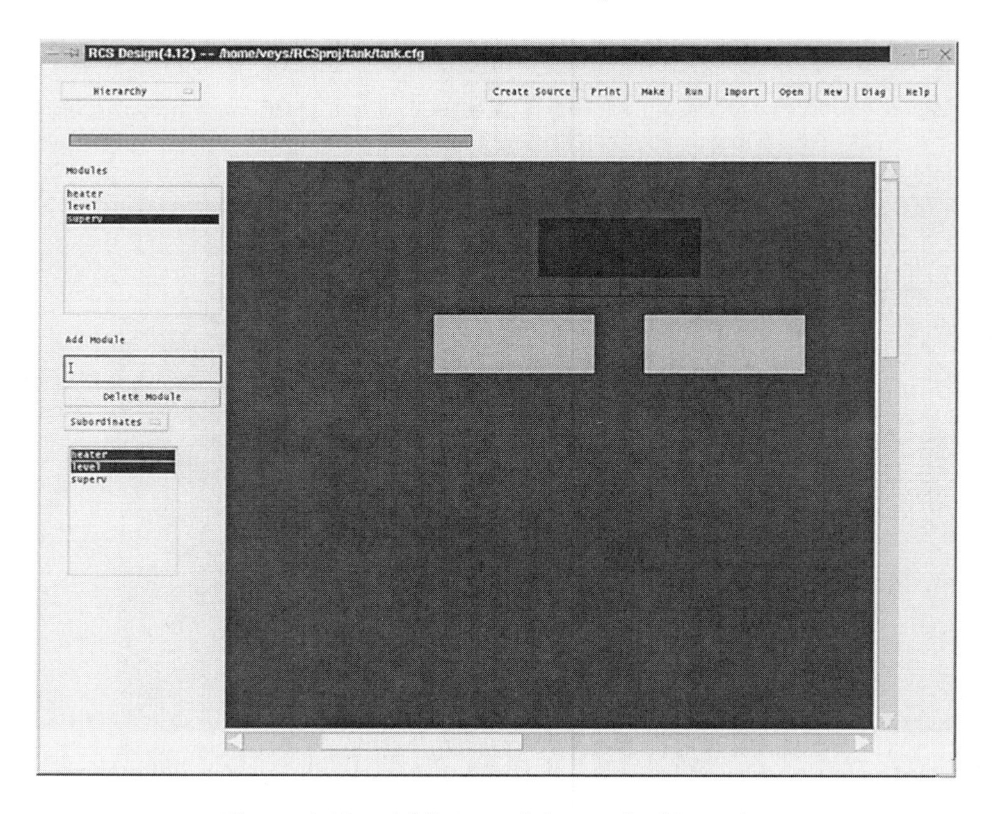

Figure 2.12: Adding modules to the hierarchy.

main part of the screen.

The box showing "Subordinates" in Figure 2.12 is a drop-down box from which we can choose other options, such as commands and auxiliary channels. We discuss auxiliary channels in Chapter 10. For now, we choose the option "Commands," which we can use to add commands to each module by selecting the module and then specifying the needed command in the "Add Command" box. All the added commands for the chosen module appear in the commands list. Figure 2.13 shows the commands for the level module in the controller for the process control problem. Recall that these were the commands that we chose during the design process (refer to Table 2.1). Note that the RCS design tool assumes that each module has, by default, commands for initializing and halting the system. Therefore, it adds the commands `INIT` and `HALT` automatically to the list of commands for each module. Then it also automatically generates a candidate implementation of the code for the corresponding functions of these commnads. If you do not need these commands, just delete them from the list of commands.

From the drop-down box on the top left corner of the screen, which shows "Hierarchy" in Figure 2.13, we can choose different views of the tool, such

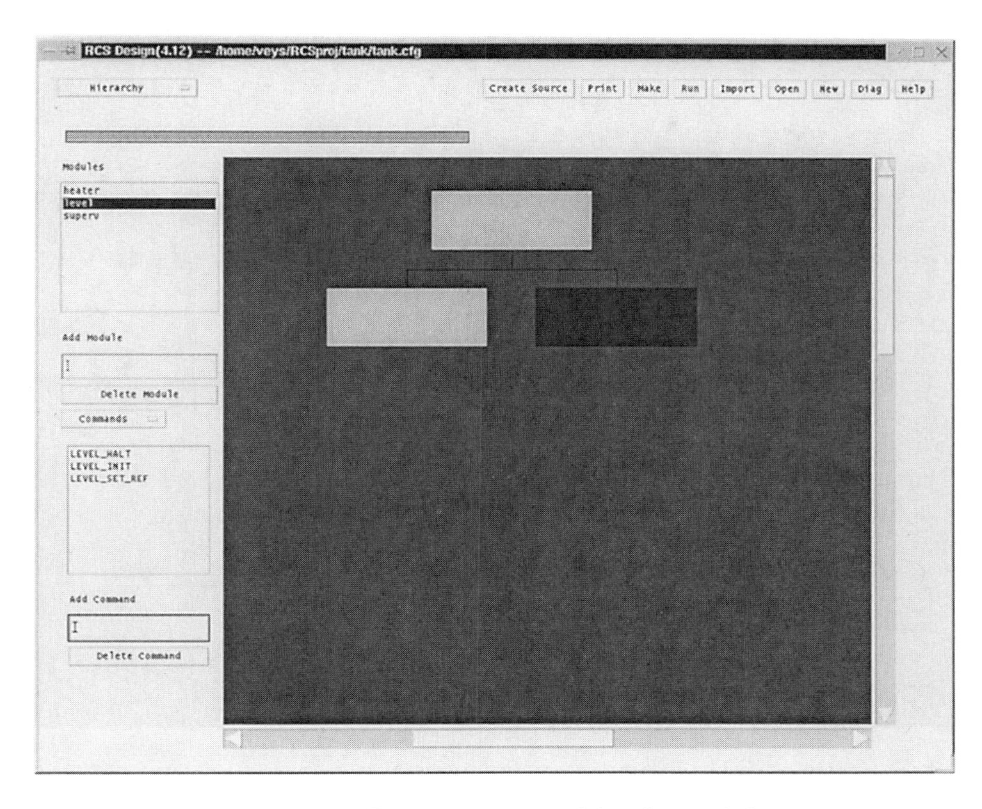

Figure 2.13: Defining command for the modules.

as "Options," "Loops/Server," "Files," and so on, and these are described as follows:

- The "Options" view is as shown in Figure 2.14. From this view we specify the various options related to the application, the platform for which we are developing the application, compile and run commands on our computer system, paths of the directories where the RCS library and the Java Development Kit are installed, whether or not to use *version control system*, and so on. We will not explain the meaning of these options here; however, the interested reader can consult Chapter 10 for details.

- From the "Loops/Servers" view, we assign all the modules to the same execution loop, called `tank`, with a cycle time of 1.0 second, as shown in Figure 2.15. (Based on the value that you specify here, an RCS timer will be established in the code. The resolution of the RCS timer depends on the system clock in the current platform; therefore, you should specify an achievable cycle time here.) Moreover, we choose only one NML server, which we call `tanksvr`, to access all the buffers in the system for the remote processes or, in other words, the diagnostics program in this application.

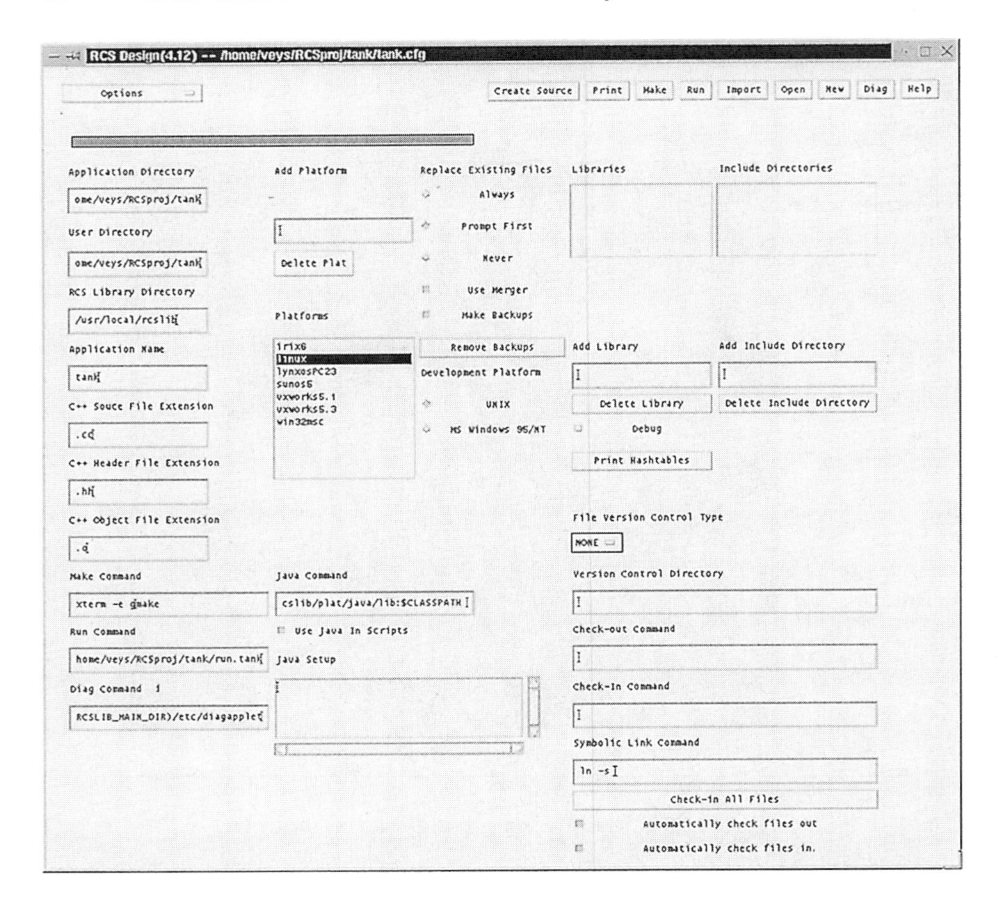

Figure 2.14: Specifying the settings needed by the application script files.

Having specified all the options needed for our application, we generate the source code for the application by pressing the "Create Source" button on the top of the tool. This also generates a particular directory structure and the script files needed for building and running the application.

- Finally, from the "Files" view we can view and edit all the files generated for the application. The "Files" view for our controller is as shown in Figure 2.16, where currently the file `heater.cc` from the list of the files on the left box is selected (shown highlighted) and the code of that file appears on the right box, from where we can edit or modify it.

2.4.2 Generated Files and Directory Structure

Recall that for the process control application we specified the name `tank` and the directory `/home/veys/RCSproj/tank`. Therefore, under the directory

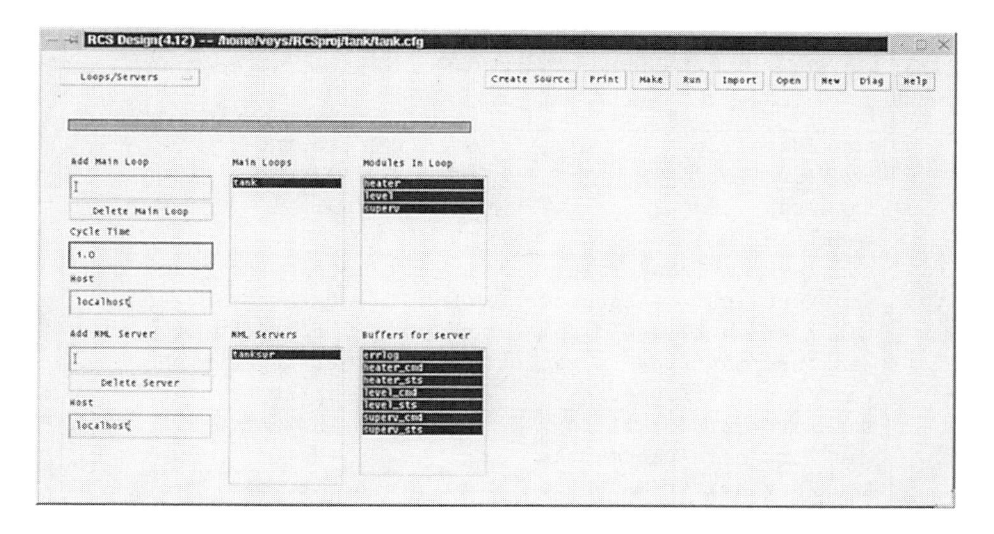

Figure 2.15: Defining the main loops and NML servers.

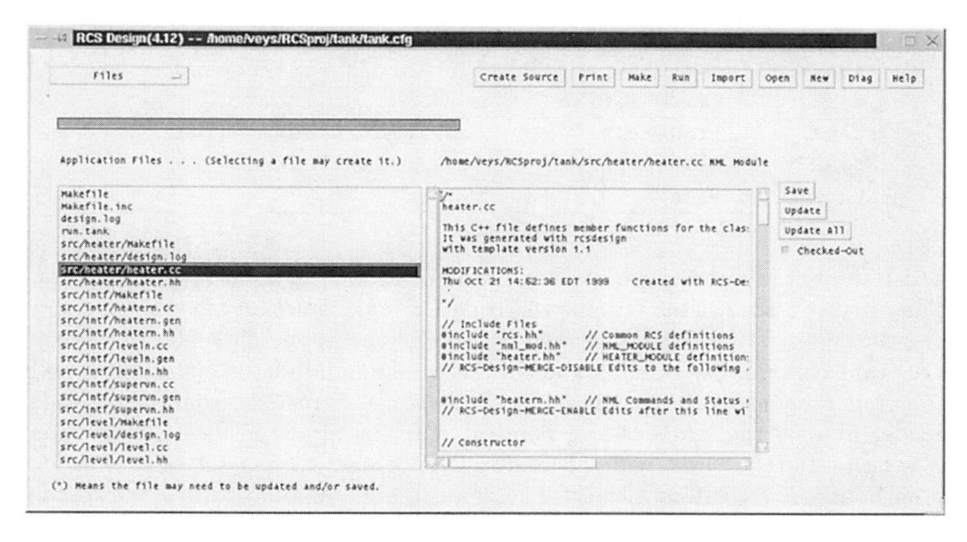

Figure 2.16: Generating the application files.

/home/veys/RCSproj/ the directory structure and files shown in Table 2.3 are generated.

The files called `Makefile` are scripts to compile or build the application, and the script `run.tank` can be used to run the application executables. The `design.log` files hold the information about the modifications that you make to the files using the RCS design tool. The files `tank.nml` and `tank.nml.local` are NML Configuration files whereas the file `tank.cfg` is an architecture file needed by the RCS diagnostics tool. The source code of the files is contained

Table 2.3: Generated files and directory structure for the process control experiment.

tank/Makefile	tank/Makefile.inc
tank/design.log	tank/run.tank
tank/tank.cfg	tank/tank.nml
tank/tank.nml.local	
tank/src/intf/Makefile	tank/src/intf/heatern.hh
tank/src/intf/heatern.cc	tank/src/intf/heatern.gen
tank/src/intf/leveln.hh	tank/src/intf/leveln.cc
tank/src/intf/leveln.gen	tank/src/intf/supervn.hh
tank/src/intf/supervn.cc	tank/src/intf/supervn.gen
tank/src/main/Makefile	tank/src/main/tankmain.cc
tank/src/main/tanksvr.cc	
tank/src/heater/Makefile	tank/src/heater/design.log
tank/src/heater/heater.hh	tank/src/heater/heater.cc
tank/src/level/Makefile	tank/src/level/design.log
tank/src/level/level.hh	tank/src/level/level.cc
tank/src/superv/Makefile	tank/src/superv/design.log
tank/src/superv/superv.hh	tank/src/superv/superv.cc
tank/src/util/Makefile	
tank/plat/linux/bin/	tank/plat/linux/include/
tank/plat/linux/lib/	tank/plat/linux/src/
tank/plat/sunos5/	...

under the `src` directory, where there is subdirectory for each of the modules. The directory `src/main` contains the code for main functions for the executables of the execution loops and NML servers in the application. The directory `src/intf` contains the code for the NML command and status messages as well as scripts for generating automatically the message source files from the message header files by the NML code generator. For instance, the file `heatern.gen` is a script for the NML code generator that is used to generate `heatern.cc` from `heatern.hh` automatically. Therefore, all the modifications done to the `heatern.hh` are also automatically reflected to the `heatern.cc`. The directory `src/util/` is empty and can be used by the programmer for other utility files needed by the application. For example, we can use this directory for the code that we use to access the DAS20 data acquisition card.

If the path for the RCS library and the Java Development Kit directories are correctly specified in the "Options" view of the design tool, then the whole application can be built for a given platform simply by typing

`make plat=plat_name` (`make plat=linux` in this case)

from the `/home/veys/RCSproj/tank/` directory. This will update any source code if there is need for any such update, and copy all the header and source files to `plat/plat_name/include/` and `plat/plat_name/src/`, respectively. Then, it will generate the object files in the `plat/plat_name/lib/` and the executables in `plat/plat_name/bin/` directories, respectively. If there are no errors and the build operation is successful, then we can run the application simply by typing `csh -f run.tank` from the `/home/veys/RCSproj/tank/` directory. However, before building and running the application, we need to modify some of the files in order to add our application-specific code. We specify these files next.

2.4.3 Files That Need To Be Modified

After generating the source code and the scripts for the application, we need to modify some of the files to add our application-specific variables and algorithms. First, we need to modify the command and status message classes in order to add the application-specific data variables. The files that contain the definition of these messages are:

```
src/intf/heatern.hh   src/intf/leveln.hh   src/intf/supervn.hh.
```

To illustrate, consider the command `HEATER_SET_REF`, which is already defined in `heatern.hh`. We need to add the data field `desired_temperature` to this message class. Similarly, to the message class `HEATER_STATUS`, we need to add the field `current_temperature`. We need to provide the data variables for the other messages of this module and the other modules in a similar way. In the following sections we provide the code for the messages and show how we add the data variables to them. Corresponding to these modifications of the message class definitions, there are some modifications that need to be done to the update functions and the constructors of the message classes in the files:

```
src/intf/heatern.cc   src/intf/leveln.cc   src/intf/supervn.cc.
```

However, as mentioned above, once we add the data variables to the message class definitions in the header files, the corresponding modifications to the source files could be done automatically using the NML code generator and the code generation scripts. (See Chapter 10 for more information on the NML code generator.) Moreover, if you are using th makefilese provided for building the application, then you do not have to worry about the source code modifications because the makefiles will take care of them automatically. We discuss the message classes and the needed update functions in a little more detail in the next section.

The other files that we need to modify for this application are

```
src/heater/heater.hh   src/level/level.hh   src/superv/superv.hh
src/heater/heater.cc   src/level/level.cc   src/superv/superv.cc
```

which contain the definition of the module classes for the three modules in the system. Our job is to add any needed extra data and function members to these classes and to "fill in" the state tables or the control algorithms within each of the command functions whose skeleton is already generated by the design tool. In the following sections, we present some of the code generated by the design tool and also show how to do the needed changes to all these files.

Another modification that we need for this application is to compile and link the code for accessing the DAS20 data acquisition card. To this end we copy all the files containing the related code to the `src/util/` directory and modify the file `src/util/Makefile` to reflect the new files in this directory. Then, we can use in our application all the functions in the code developed previously. We will not discuss this code here, as it is specific only to our particular experiment and the emphasis here is on the RCS development.

Once we are done with the modifications above we are ready to build and test our RCS controller for the process control experiment.

2.5 Code and Operation of the Process Control System

In this section we look at some of the source code files and scripts for the process control experiment in more detail. This will help the reader to get an idea of how RCS programming proceeds and will help clarify the operation of an RCS controller. This code can also serve as an example for beginning RCS library users.

2.5.1 NML Configuration File for the Tank Experiment

NML configuration files are text files that are used to specify which buffers and processes are used in the application. They can be generated automatically by the RCS design tool. However, sometimes the user may need to modify them manually. This is because they are provided so that the user can change buffer sizes, communication protocols, mutual exclusion types, buffer locations, and so on, without a need to recompile or rebuild the application. We discuss NML configuration files in detail in Chapter 7.

Before proceeding with an explanation of the configuration file for the process control application, recall the buffers and processes in our application (refer to Figure 2.9). In addition to command and status buffers, in our application we have one more buffer that is used for error messages to be shared by all the processes (not shown in Figure 2.9). In Example 2.1 we provide the NML configuration file for the tank experiment (i.e., the file `tank/tank.nml`, which was generated by the design tool, and we modified it a little as we describe below). The lines starting with "`B`" are the lines defining the buffers, the lines starting with "`P`" are the ones defining the processes, and the "`#`" symbol specifies the comment lines.

Example 2.1: NML configuration file for the tank

```
# NML Configuration file for the tank application

# Buffers
# Name          Type    Host  size neut? RPC#     buf# MP  . . .
```

```
B heater_cmd  SHMEM  eepc100 128  0  0x2010f174  1   4  8060 STCP=6001 disp
B heater_sts  SHMEM  eepc100 128  0  0x2010f174  2   4  8061 STCP=6001 disp
B level_cmd   SHMEM  eepc100 128  0  0x2010f174  3   4  8062 STCP=6001 disp
B level_sts   SHMEM  eepc100 128  0  0x2010f174  4   4  8063 STCP=6001 disp
B superv_cmd  SHMEM  eepc100 128  0  0x2010f174  5   4  8064 STCP=6001 disp
B superv_sts  SHMEM  eepc100 128  0  0x2010f174  6   4  8065 STCP=6001 disp
B errlog      SHMEM  eepc100 256  0  0x2010f174  7   4  8066 STCP=6001 disp queue

# Processes
# Name        Buffer         Type    Host     Ops server? timeout master? cnum

# heater(0)
P heater     heater_cmd     LOCAL  eepc100   R    0       10.0      0      0 poll
P heater     heater_sts     LOCAL  eepc100   W    0       10.0      0      0
P heater     errlog         LOCAL  eepc100   RW   0       10.0      0      0

# level(1)
P level      level_cmd      LOCAL  eepc100   R    0       10.0      0      1 poll
P level      level_sts      LOCAL  eepc100   W    0       10.0      0      1
P level      errlog         LOCAL  eepc100   RW   0       10.0      0      1

# superv(2)
P superv     superv_cmd     LOCAL  eepc100   R    0       10.0      0      2 poll
P superv     superv_sts     LOCAL  eepc100   W    0       10.0      0      2
P superv     heater_cmd     LOCAL  eepc100   W    0       10.0      0      2
P superv     heater_sts     LOCAL  eepc100   R    0       10.0      0      2
P superv     level_cmd      LOCAL  eepc100   W    0       10.0      0      2
P superv     level_sts      LOCAL  eepc100   R    0       10.0      0      2
P superv     errlog         LOCAL  eepc100   RW   0       10.0      0      2

# tanksvr(3)
P tanksvr    errlog         LOCAL  eepc100   RW    1       INF      1      3
P tanksvr    superv_cmd     LOCAL  eepc100   RW    1       INF      1      3
P tanksvr    superv_sts     LOCAL  eepc100   RW    1       INF      1      3
P tanksvr    heater_cmd     LOCAL  eepc100   RW    1       INF      1      3
P tanksvr    heater_sts     LOCAL  eepc100   RW    1       INF      1      3
P tanksvr    level_cmd      LOCAL  eepc100   RW    1       INF      1      3
P tanksvr    level_sts      LOCAL  eepc100   RW    1       INF      1      3
```

Buffers

In this example, all the buffers are located on the host called `eepc100`, which is the Linux station. Since all the buffers will be accessed, both by some remote and local processes, they are declared as "`SHMEM`," which means *shared memory*. The buffers, excluding `errlog`, are of size 128 bytes, and these buffers will not be encoded in machine-independent format because `neut` is set to zero. All the buffers are given unique buffer numbers, where `heater_cmd` is number 1 and `errlog` is number 7. The `RPC#` specifies the *remote procedure call* (RPC) number in case RPC is used for remote communications. This field is kept here for backword compatiblity with the older versions of the RCS library and is not needed anymore. The maximum procedure (`MP`) specifies the maximum number of processes that will access this buffer, and it is set to 4 for all the buffers. This is because as a convention, the design tool sets this number to the number of all the processes in the system. The user can modify this number and specify the actual number of the processes that will access this buffer. For example, the supervisor buffers will be accessed by the supervisor itself and the NML servers; therefore the maximum procedure number for these can be set to 2. Similarly, the buffers for the level and heater module are expected to be accessed by the level and heater modules, respectively, plus the supervisor and NML servers. For this reason, their maximum procedure number can be set to 3, and since all the procedures can write to the `errlog` buffer, the number of procedures to it should remain as 4. We choose the `STCP` (Simplified Transmission Control Protocol) communication protocol for this experiment; however, you can equally well chose another protocol. For some applications that need faster communication, you may need to choose `UDP` (User Datagram Protocol) because it is faster then `TCP` (Transmission Control Protocol). This is not the case here because all the time-critical processes communicate through the buffers locally, since they are on the same computer, which does not slow them down. The only network communication needed here is the communication with the diagnostics tool (i.e., the operator interface module). The port number for the communication protocol is chosen to be 6001 and the buffers are written in displayable format. This is because of the added keyword `disp`. Note that `errlog` is defined as `queue`, implying that new messages will not erase the old ones, since all the processes will use it for their error messages and we do not want one process to overwrite the message of other processes.

From the original configuration file generated by the design tool in the buffer lines we changed the size of the buffers, the communication protocol from `TCP` to `STCP`, the TCP port number, and the format of the data in the messages from `xdr` (eXternal Data Representation) to `disp`.

Processes

Since all the processes (except the diagnostic one that is located on the interface computer) are running on the Linux PC, they are declared as `LOCAL` to the buffers, which are also on the same PC. As mentioned before, if we were developing this application under DOS, we would not be able to lay out the buffers and processes as we did. In that case, the buffers would be on `eepc99` and the `superv`, `level`, and `heater` processes would be defined as `REMOTE` to the buffers because they would be running on `eepc100`. In contrast, the NML server, called `tanksvr` (i.e., the process that runs all the six NML servers in Figure 2.9), would still be defined as `LOCAL` because it would be on the same PC with the buffers.

In the NML file it is required that every process has an entry with the buffer it will access. Therefore, you define the same process multiple times with each buffer that it will access. Note that every process has an entry together with the `errlog` buffer. The process called `tanksvr`, i.e., the NML server for the buffers, is set as a server and master for the buffers by assigning 1 to the `server` and `master` fields, respectively. In general, we can equally well choose other processes to be master for a buffer, however, it is a good practice to choose the NML server as master. This is because the master is responsible for creating the buffers and therefore it should be run first, however, sometimes it may not be a good idea to run the module processes before the communication with the rest of the system is established. The connection number (i.e., `cnum`) specifies the number under which the process conects to a particular buffer. This number should be between 0 and the `MP` number set for this buffer, minus one. Each process connects with a different number to a given buffer. For instance, the `cnum` for the `tanksvr` to all the buffers is 3. As a convention, the design tool assigns a number to each process (see the comment line for each process) and assigns this number as a connection number of this process to all the buffers it accesses. This is also the reason why it assigns `MP` to be the number of all the processes in the system. To illustrate, assume that we set the `MP` number of `superv_cmd` buffer to 2. Then we will have only two processes, (i.e., `superv` and `tanksvr`) connecting to it with `cnum` 0 and 1, respectively (i.e., if we modify the `MP` number of a buffer, we may have to also modify the connection numbers of the processes connecting to it). The string `INF` under field `timeout` means that there is no timeout for the given process while waiting to access the buffer, and 10.0 sets up a timeout of 10 seconds. The `Ops` (i.e., the options) field specifies the read and write permission of the process to the buffer. You can note that, in general, the processes have "`R`" (i.e., read) access to its command buffer and "`W`" (i.e., write) access to its status buffer, and the NML server has both read and write permission to the buffers it accesses. The string `poll` is added if you need the process to poll the buffer continuously for a new message.

From the original configuration file generated by the design tool in the process lines we changed the `Ops` for some of the buffers since the design tool by default generates all the processes to connect as `RW` to the buffers they access. We aslo changed the timeout of the server to `INF` and added the `poll` string to some of the lines. We did these modifications because we want each process to poll continuously for new commands, and we want the NML server to wait for the other processes as long as they are accessing the buffers.

Note that once this file is generated by the RCS design tool you may need to do some simple modifications based on the needs of your system. For instance, if the TCP (or UDP) port number generated is already in use by some other process, then you may need to modify the port number for your RCS application in order not to have conflict with the other processes.

2.5.2 Command and Status Message Classes for the Tank

In RCS the communications between the processes occur via NML messages. Recall that during the design process of the controller we defined the command and status messages based on the task decomposition analysis and the controller hierarchy chosen. A command message was a message that requests an action to be performed by a module, whereas a status message is a message that provides information about the current conditions or state of the module. Refer to Table 2.1 for a list of the messages in the tank application. Messages are application specific; therefore, we need to develop and code them. From the preceding section we know that we can define the messages for a particular application through the RCS design tool and generate the skeleton of their code automatically. However, we still need to add the application-specific data variables to these messages. As mentioned before, for example, we have to add a variable for holding the current temperature in the heater modules' status message and a variable for the desired temperature to the heater modules' set reference message.

In the preceding section we generated the code for the process control application, including the code for the messages. Example 2.2 shows the code for the messages for the level module. We show the definition of only one status and one command message. Since the definition of the other messages is the same, we omit them. This code is part of the header file `src/intf/leveln.hh`. Note that the code already reflects the necessary modifications (i.e., the variables `double current_level` and `double desired_level` are already added to the corresponding messages after the comment line `// Place custom variables here`).

Example 2.2: Level command and status message classes

```
/*
leveln.hh
This C++ header file defines the NML Messages used for command
and status by LEVEL_MODULE
*/

// Prevent Multiple Inclusion
#ifndef LEVELN_HH
#define LEVELN_HH

// Include Files
#include "rcs.hh"  // Common RCS definitions

// Define the integer type ids.
#define LEVEL_STATUS_TYPE 3000
#define LEVEL_HALT_TYPE 3001
#define LEVEL_INIT_TYPE 3002
#define LEVEL_SET_REF_TYPE 3003

// Define the NML Message Classes
// Status Class
class LEVEL_STATUS : public RCS_STAT_MSG
{
public:
  // Normal Constructor
  LEVEL_STATUS();

  // CMS Update Function
  void update(CMS *);

  // Place custom variables here.
  double current_level;
};
```

```
// Command Classes
// ...
// The classes for LEVEL_INIT and LEVEL_HALT commands
// ...
class LEVEL_SET_REF : public RCS_CMD_MSG
{
public:
  //Constructor
  LEVEL_SET_REF();

  // CMS Update Function
  void update(CMS *);

  // Place custom variables here.
  double desired_level;
};

// Declare NML format function
extern int levelFormat(NMLTYPE, void *, CMS *);

#endif  // LEVELN_HH
```

Note that before defining a particular message you need to define a unique message type specifier with _TYPE added to the message. For example, if you want to define a message called LEVEL_SET_REF, you have to define the type identifier LEVEL_SET_REF_TYPE for the message using something like #define LEVEL_SET_REF_TYPE 3003, as shown in the example above. The number that you assign to that identifier (3003 above) should be unique within this application. This will be used for distinguishing this message from the other messages in the application. Note that the RCS design tool automatically generates all the type definitions with unique numbers.

You will notice immediately that the status message is derived from the RCS_STAT_MSG class and the command messages are derived from the RCS_CMD_MSG class. RCS_CMD_MSG and RCS_STAT_MSG are predefined RCS command and status message base classes provided by the RCS library. They are derived from the NMLmsg base class and carry command or status message-specific fields or data members.

Another important issue to mention here is that every message has its own CMS update function, which should be called for any new data field that is added to the message class. CMS update functions are used basically to encode or decode the message appropriately in the buffer. To illustrate, recall that a field called current_level was added to the LEVEL_STATUS message class; therefore, in the implementation of the update function of that class, a statement for updating this field should be present.

This is shown in Example 2.3, which is an extract from the source file src/intf/leveln.cc and can be totally generated by the NML code generator (i.e., we do not have to add any code to this file if we use the automatic code generation tools of the RCS library) for the LEVEL_STATUS and LEVEL_SET_REF messages. Example 2.3 also shows the constructors for these two messages.

Example 2.3: Constructors and update functions of the messages

```
//  Constructor for LEVEL_SET_REF
LEVEL_SET_REF::LEVEL_SET_REF()
  : RCS_CMD_MSG(LEVEL_SET_REF_TYPE,sizeof(LEVEL_SET_REF))
{
  desired_level = (double) 0;
}

//  NML/CMS Update function for LEVEL_SET_REF
void LEVEL_SET_REF::update(CMS *cms)
{
  cms->update(desired_level);
}


//  Constructor for LEVEL_STATUS
LEVEL_STATUS::LEVEL_STATUS()
  : RCS_STAT_MSG(LEVEL_STATUS_TYPE,sizeof(LEVEL_STATUS))
{
  current_level = (double) 0;
}

//  NML/CMS Update function for LEVEL_STATUS
void LEVEL_STATUS::update(CMS *cms)
{
  cms->update(current_level);
}
```

Note that the constructors basically call the constructor of the base class and initialize the new data variables. In other words, the constructor of the LEVEL_SET_REF class calls the constructor of its base class RCS_CMD_MSG through the statement

```
RCS_CMD_MSG(LEVEL_SET_REF_TYPE, sizeof(LEVEL_SET_REF))
```

and the constructor of the LEVEL_STATUS class calls the constructor of its base class RCS_STAT_MSG through the

```
RCS_STAT_MSG(LEVEL_STATUS_TYPE,sizeof(LEVEL_STATUS))
```

statement. The base constructors allocate the needed memory for the message and assign the type variable. The update function, on the other hand, calls the CMS update functions to perform the necessary update operation. If a message class does not have extra fields from the predefined message classes, then their constructors and update functions are empty.

To better understand the operation of the update function, recall the structure of the single process shown in Figure 2.10. From this figure you can see where to find update functions within the operational structure of a process. Note from the figure that we also need to have a format function. Example 2.4 shows a function called `levelFormat` that is the format function for level module messages. Note that this function also is an exerpt from the file src/intf/leveln.cc, which is generated totally by the design tool and the NML code generator, and it was first declared in the header file shown in Example 2.2. It is a simple switch statement that checks the incoming message and then calls the update function corresponding to this message.

Example 2.4: Format function for the level module messages

```
/*
* NML/CMS Format function : levelFormat
*/
int levelFormat(NMLTYPE type, void *buffer, CMS *cms)
{
  switch(type)
    {
    case LEVEL_HALT_TYPE:
      ((LEVEL_HALT *) buffer)->update(cms);
      break;

    case LEVEL_INIT_TYPE:
      ((LEVEL_INIT *) buffer)->update(cms);
      break;

    case LEVEL_SET_REF_TYPE:
      ((LEVEL_SET_REF *) buffer)->update(cms);
      break;

    case LEVEL_STATUS_TYPE:
      ((LEVEL_STATUS *) buffer)->update(cms);
      break;

    // unknown type
    default:
      return(0);      // 0 signifies didn't find
    }
  return 1;           // 1 signifies found it
}
```

As we stated before, the header files for the message classes can be generated by the RCS design tool. Moreover, given the header file, the format function, the update functions, and other member functions of the messages can be generated using the NML code generator (which is also integrated with the design tool). Here we present this code simply to help the beginner programmer to better understand the underlying code and its operation.

2.5.3 Module Classes

In RCS applications the programmer first creates a general description of the system through the task analysis and then decides how many buffers, controller modules, and message classes are needed. Then we need to write (or simply generate via the design tool) the code for these. Above, we described the code for the NML configuration file, and the command and status messages, together with the format and update functions. In this section we describe the code of the control nodes (NML modules). Note that the development of all these parts of the application code is not totally independent because they share information. For example, in the initialization of the NML servers you have to use the names of the buffers as defined in the NML configuration file. Moreover, most of the function members of the modules take as input parameters the pointers to the corresponding message classes. Therefore, use of the RCS design tool makes the job of the programmer easier, because it generates the skeleton of the code in a consistent manner. Therefore, the job left to the programmer is only to "fill in" some parts of the code. For the code of the modules we need to develop the application-specific control or estimation routines. These are expressed as state tables for the commands in the application, and inserted in the functions corresponding to these commands.

Definition of an NML Module

Example 2.5, which is the file `src/level/level.hh` generated by the design tool with a few modifications, shows the code for the definition of the level module in the tank application.

Example 2.5: Control module for the level

```
/*
level.hh --- This C++ header file defines the class LEVEL_MODULE
*/

// Prevent Multiple Inclusion
#ifndef LEVEL_HH
#define LEVEL_HH

// Include Files
#include "rcs.hh"  // Common RCS definitions
#include "nml_mod.hh"  // NML_MODULE definitions
#include "leveln.hh"  // NML Commands and Status definitions for level

#include "levelclass.hh" // LevelClass definitions

class LEVEL_MODULE: public NML_MODULE, public LevelClass
{
public:
  LEVEL_MODULE(); // Constructor

  // Overloaded Functions
  void PRE_PROCESS();
  void DECISION_PROCESS();
  void POST_PROCESS();
  void INITIALIZE_NML();

  // Command Functions
  void INIT(LEVEL_INIT *);
  void HALT(LEVEL_HALT *);
  void SET_REF(LEVEL_SET_REF *);
```

```
  // Convenience Variables
  LEVEL_STATUS *level_status;

private:
  // Add custom variables and functions here.
  double desired_level;
  double maxLevel;
  double minLevel;
};

#endif  // LEVEL_HH
```

Note that the class `LEVEL_MODULE` is derived from both `NML_MODULE` and `LevelClass`. The class `NML_MODULE` is a generic class in the RCS library (i.e., it serves as a base class for implementing a control module in the RCS library). On the other hand, the class `LevelClass` is a class that we created for the functions and the variables of the pumps and the level sensor and the DAS20 acquisition board interface. Note that we added some data members, such as `desired_level`, `maxLevel`, and `minLevel`, to the class. These are convenience variables to hold the desired level received by the last command, the maximum allowable safety level, and the minimum allowable safety level, respectively. The member functions `DECISION_PROCESS()`, `PRE_PROCESS()`, and `POST_PROCESS()` are virtual functions of the class `NML_MODULE`. Recall that we first mentioned these functions while introducing the RCS control node (refer to Figure 2.2).

NML Module Constructor

The constructors are functions where memory is allocated and the variables are initialized (see Appendix A). The constructor of the level module creates an NML object and initializes an NML communication channel. This is done by calling the NML constructors. The process and buffer names passed to the NML constructor should be as defined in the NML configuration file. The code in Example 2.6, which is a part of the file `src/level/level.cc` generated by the design tool, shows the constructor of the level module.

Example 2.6: Constructor of the level module

```
// Constructor
LEVEL_MODULE::LEVEL_MODULE()
{
  // Initialize the NML channels
  setErrorLogChannel(new NML(nmlErrorFormat, "errlog", "level", "tank.nml"));
  setCmdChannel(new RCS_CMD_CHANNEL(levelFormat, "level_cmd", "level",
    "tank.nml"));
  level_status = new LEVEL_STATUS();
  setStatChannel(new RCS_STAT_CHANNEL(levelFormat, "level_sts", "level",
      "tank.nml"), level_status);

  // Add additional code to initialize the module here.
  minLevel = MIN_SAFETY_LEVEL;
  maxLevel = MAX_SAFETY_LEVEL;
```

```
  GetLevel();
  CopyLevel();
}
```

The function setErrorLogChannel establishes an NML communication channel of this module to the errlog buffer, whereas the functions setCmdChannel and setStatChannel, which are member functions of the NML_MODULE, initialize NML channels to the command and status buffers of the module. Note that the names of the buffers and processes are the same as those specified in the NML configuration file in Example 2.1. Moreover, the name of this file, tank.nml is passed to the constructor so that during the initialization time the constructor can read from the NML file which communication protocols, mutual exclusion techniques, and so on, to use.

Note also that we added some extra code after the comment line // Add additional code to initialize the module here to initalize the variables minLevel and maxLevel, and to read the liquid level in the tank at startup. The constants MIN_SAFETY_LEVEL and MAX_SAFETY_LEVEL specify the safety levels and are defined within the header file levelclass.hh included in level.hh in Example 2.5.

Since the level module is on the lowest level of our controller hierarchy, it does not have any subordinates. Therefore, in the code in Example 2.6 there is no initialization of communication channels to any subordinates. Example 2.7 shows the constructor of the supervisor module.

Example 2.7: Constructor of the supervisor module

```
// Constructor
SUPERV_MODULE::SUPERV_MODULE()
{
  // Initialize the NML channels
  setErrorLogChannel(new NML(nmlErrorFormat, "errlog", "superv", "tank.nml"));
  setCmdChannel(new RCS_CMD_CHANNEL(supervFormat, "superv_cmd", "superv",
                                    "tank.nml"));
  superv_status = new SUPERV_STATUS();
  setStatChannel(new RCS_STAT_CHANNEL(supervFormat, "superv_sts", "superv",
                                      "tank.nml"), superv_status);
  // Initialize the NML channels to subordinates
  heater_sub_num = addSubordinate(
       new RCS_CMD_CHANNEL(heaterFormat, "heater_cmd", "superv", "tank.nml"),
       new RCS_STAT_CHANNEL(heaterFormat, "heater_sts", "superv", "tank.nml"));
  heater_status = (HEATER_STATUS *) statusInData[heater_sub_num];

  level_sub_num = addSubordinate(
       new RCS_CMD_CHANNEL(levelFormat, "level_cmd", "superv", "tank.nml"),
       new RCS_STAT_CHANNEL(levelFormat, "level_sts", "superv", "tank.nml"));
  level_status = (LEVEL_STATUS *) statusInData[level_sub_num];

  // Add additional code to initialize the module here.
  ...
}
```

Note first, how similar the code in Example 2.7 is to the one in Example 2.6. This shows the modularity of the code in RCS. The major difference between the code in the two constructors, excluding the difference between the specific module-dependent code, is that the supervisor constructor has extra code for initializing communication channels to its subordinates. This is done through the function `addSubordinate`, which is a member function of the `NML_MODULE` class. This function returns a unique identification (assigned to the variables `heater_sub_num` and `level_sub_num` in the code above), which can be used to access the status information, located at `statusInData[sub_num]`, of the corresponding subordinate, or to send a command to it. Note also that for conveniance the two variables `level_status` and `heater_status`, pointing to `statusInData[level_sub_num]` and `statusInData[heater_sub_num]`, respectively, are defined. These variables can be used to directly access the status information of the subordinate modules. How the `sub_num` is used for sending commands to the subordinates is shown in Example 2.11.

`DECISION_PROCESS()`, `PRE_PROCESS()`, and `POST_PROCESS()` functions

Recall that the function `DECISION_PROCESS()` is the function where the current situation is evaluated, and based on the user inputs, commands from the supervisor, internal state of the module, and status of the subordinates, the appropriate decision is made and the function for the action desired is called. In its simplest form it can be implemented in a very similar way to the NML format function. In other words, it can be implemented as a simple `switch` statement that calls the command function of the currently received command, and that command is passed as a parameter to the called function. In this simple form, it is generated by the RCS design tool. Example 2.8, which is an exerpt from the file `src/level/level.cc` generated by the design tool, shows this implementation. We use this simple implementation for the level module, since we do not need more sophisticated logic for the decision process. Note, however, that the user may implement this function in another way based on the needs of the application in hand. Note also that the `DECISION_PROCESS()` function should be implemented and it should call one of the command functions based on some logic.

Example 2.8: `DECISION_PROCESS()` for the level module

```
/*
DECISION_PROCESS

The DECISION_PROCESS function is called every cycle as long as there is a
non-zero command.
It is expected to call a command function based on commandInData->type.
*/
void LEVEL_MODULE::DECISION_PROCESS()
{
  switch(commandInData->type)
    {
```

```
    case LEVEL_INIT_TYPE:
      INIT((LEVEL_INIT *)commandInData);
      break;

    case LEVEL_HALT_TYPE:
      HALT((LEVEL_HALT *)commandInData);
      break;

    case LEVEL_SET_REF_TYPE:
      SET_REF((LEVEL_SET_REF *)commandInData);
      break;

    default:
      logError("The command %d is not recognized.",commandInData->type);
      break;
    }
}
```

You do not necessarily need to implement the other two functions (i.e., the virtual functions PRE_PROCESS() and POST_PROCESS()). However, if there is a need for something to be done immediately before calling the controller or immediately after performing the control operation, such as reading a value from the plant or saving some value to a variable or file or outputting something to the screen, and so on, then the programmer can define and implement these. The implementation of these in our example is as shown in Example 2.9.

Example 2.9: Level module PRE_PROCESS() and POST_PROCESS()

```
/*
PRE_PROCESS

The PRE_PROCESS function is called every cycle after the command and
subordinates status have been read but before DECISION_PROCESS is called.
It is intended to be used for tasks such as sensory processing that should
be performed every cycle regardless of the current command or state.
*/
void LEVEL_MODULE::PRE_PROCESS()
{
  // Measure the level
  GetLevel();

  // Unless later specified else the pumps are off
  Pumper[HOT_TANK] = OFF;
  Pumper[REACT_TANK] = OFF;
}

/*
POST_PROCESS

The POST_PROCESS function is called every cycle after DECISION_PROCESS is
called but before the status and the subordinates commands  have been written.
It is intended to be used for tasks such as output filters that should
be performed every cycle regardless of the current command or state.
*/
```

```
void LEVEL_MODULE::POST_PROCESS()
{
  // Ouput the control values
  Pump(HOT_TANK, Pumper[HOT_TANK]);
  Pump(REACT_TANK, Pumper[REACT_TANK]);

  // Save and report the information
  CopyLevel();
  level_status.level = level1[REACT_TANK];
  rcs_print("Reference Level: %.2lf ",maxLevel);
  rcs_print("Level: %.2lf \n",level1[REACT_TANK]);
}
```

The functions GetLevel(), CopyLevel(), and Pump(tank #,status) and the array variables level1 and Pumper are members of the class LevelClass. The function Pump(tank #,status) outputs the specified status, either ON or OFF, to the specified tank, whereas the functions GetLevel() and CopyLevel() read the level from the physical plant and save the previous level for control purposes. We use the array level1 to hold the level in both of the tanks (for this experiment we use only the value of the level in the reaction tank), and the array Pumper is a variable to specify the control values for the pumps. Note that the function GetLevel() performs all the necessary conversion from the voltage read to the value of the level in the tank. As discussed before, we will not go into the details of these conversions since they are specific to our laboratory experiment.

As you can see from the code, in the PRE_PROCESS() function we perform the sensory processing (i.e, measure the level of the value and perform the necessary conversions). In the POST_PROCESS() function, on the other hand, we output the control to the plant, save some of the useful values, update the status message to be passed to the supervisor or the operator for diagnostics, and print some information to the screen. These are simple examples of how you can use these functions. In these functions you can also calculate the derivative or integral of the error if the controller needs them. The controller here is of the ON/OFF type, so there is no need for such calculations.

Command Function Implementation

As we saw above in the DECISION_PROCESS() function, one of the command functions is called based on the current command received or some other logic. In the implementation in Example 2.8, if the command LEVEL_SET_REF is received, then the member function SET_REF is called. The implementation of this function is application-specific; therefore, it should be developed by the RCS programmer. Basically, the RCS programmer has to convert the state table for this command (refer to Table 2.2) into code and insert it in the body of the function generated by the design tool. Example 2.10 shows the simplest implementation of the state table for the LEVEL_SET_REF command, or in other words, the implementation of the SET_REF function.

Example 2.10: Implementation of the LEVEL_SET_REF command

```
/*
  SET_REF
  Parameter(s): LEVEL_SET_REF *cmd_in -- NML Message sent from superior.
  */
void LEVEL_MODULE::SET_REF(LEVEL_SET_REF *cmd_in)
{
  // Put state table for LEVEL_SET_REF here.
  if (STATE_MATCH(NEW_COMMAND))
    {
      if ((level1[REACT_TANK] > maxLevel) || (level1[REACT_TANK] < minLevel))
        {
          status = RCS_ERROR;
          stateNext(S4);
        }
      else
        {
          desired_level = cmd_in->level;
          if (desired_level > maxLevel)
            {
              rcs_print("Desired level > Max allowable level!!! \n");

              desired_level = maxLevel;
              rcs_print("Desired level set to %f \n", desired_level);
            }
          else if (desired_level < minLevel)
            {
              rcs_print("Desired level < Min allowable level!!! \n");

              desired_level = minLevel;
              rcs_print("Desired level set to %f \n", desired_level);
            }
          rcs_print("Reference level set to %f \n", desired_level);

          if (level1[REACT_TANK] > desired_level)
            {
              status = RCS_EXEC;
              stateNext(S1);
            }
          else if (level1[REACT_TANK] < desired_level)
            {
              status = RCS_EXEC;
              stateNext(S2);
            }
          else
            {
              status = RCS_DONE;
              stateNext(S3);
            }
        }
    }
  else if (STATE_MATCH(S1))
    {
      // Fill State
      if ((level1[REACT_TANK] > maxLevel) || (level1[REACT_TANK] < minLevel))
        {
```

```
          Pumper[HOT_TANK] = OFF;
          Pumper[REACT_TANK] = OFF;
          status = RCS_ERROR;
          stateNext(S4);
        }
      else if (level1[REACT_TANK] < desired_level)
        {
          Pumper[HOT_TANK] = ON;
        }
      else if (level1[REACT_TANK] > desired_level)
        {
          Pumper[HOT_TANK] = OFF;
          stateNext(S2);
        }
      else
        {
          Pumper[HOT_TANK] = OFF;
          Pumper[REACT_TANK] = OFF;
          status = RCS_DONE;
          stateNext(S3);
        }
    }
  else if (STATE_MATCH(S2))
    {
      // Empty State
      if ((level1[REACT_TANK] > maxLevel) || (level1[REACT_TANK] < minLevel))
        {
          Pumper[HOT_TANK] = OFF;
          Pumper[REACT_TANK] = OFF;
          status = RCS_ERROR;
          stateNext(S4);
        }
      else if (level1[REACT_TANK] > desired_level)
        {
          Pumper[REACT_TANK] = ON;
        }
      else if (level1[REACT_TANK] < desired_level)
        {
          Pumper[REACT_TANK] = OFF;
          stateNext(S1);
        }
      else
{
          Pumper[HOT_TANK] = OFF;
          Pumper[REACT_TANK] = OFF;
          status = RCS_DONE;
          stateNext(S3);
        }
    }
  else if(STATE_MATCH(S3))
    {
      // Idle State
      Pumper[HOT_TANK] = OFF;
      Pumper[REACT_TANK] = OFF;
      status = RCS_DONE;

      if ((level1[REACT_TANK] > maxLevel) || (level1[REACT_TANK] < minLevel))
```

```
          {
            stateNext(S4);
          }
        else if (level1[REACT_TANK] > desired_level)
          {
            status = RCS_EXEC;
            stateNext(S1);
          }
        else if (level1[REACT_TANK] < desired_level)
          {
            status = RCS_EXEC;
            stateNext(S2);
          }
      }
    else if(STATE_MATCH(S4))
      {
        // Error State
        Pumper[HOT_TANK] = OFF;
        Pumper[REACT_TANK] = OFF;
        status = RCS_ERROR;

        rcs_print("LEVEL_MODULE---LEVEL_SET_REF: ERROR!!!");
        if (level1[REACT_TANK] > maxLevel)
          {
            rcs_print("Current level > Max allowable level!!! \n");
          }
        else if (level1[REACT_TANK] < minLevel)
          {
            rcs_print("Current level < Min allowable level!!! \n");
          }
      }
}
```

We will not discuss this code in detail since by simple inspection the user can see that it is the implementation of the state table in Table 2.2. Basically, the code assigns the values of the control outputs to the pumps through the variable `Pumper`. Recall that these values are later output to the plant within the `POST_PROCESS` function. As a difference, there is only a new state called `NEW_COMMAND`. Whenever a new command is received, the operation always starts from this state. This is a convention used in the RCS library. Whenever, a new command is received, the state is reset to `NEW_COMMAND` automatically, so that the operation always starts from a known state. It can be used to perform any needed initializations before proceeding with the execution of the task.

The reader should keep in mind that this is very simple example implementation, used just to illustrate the basic ideas of using the RCS library. Clearly, you can implement better and more efficient procedures. For instance, with this implementation when the reference level is reached, the control algorithm will start switching between the two states "fill" and "empty" since we will always have some noise in the measurements. This could be avoided by using a small dead-band around the desired level. Moreover, if for some reason an error occurs, such as going little below the minimum level, the algorithm will stop and report an error. Instead, it could be designed to report an error and also to try to correct it. In other words, in this case it could try increase the level. Nevertheless, the example serves as an illustration of a typical state table.

Recall that during execution of a command function the module can send commands to its subordinates. Since the level module does not have any subordinates, it does not send any commands.

Example 2.11 shows a portion of the INIT function of the supervisor. It is called when the command SUPERV_INIT is received by the supervisor. The code in the example is as it was generated by the RCS design tool before our modifications. As mentioned before, the RCS design tool assumes that every module has functions for initialization and halting the system and generates example code for them, unless the user deletes these commands from the commands list.

Example 2.11: Implementation of the SUPERV_INIT command

```
/*
  INIT
  Parameter(s): SUPERV_INIT *cmd_in -- NML Message sent from superior.
  */
void SUPERV_MODULE::INIT(SUPERV_INIT *cmd_in)
{
  HEATER_INIT heaterInitMsg;
  LEVEL_INIT levelInitMsg;

  if(STATE_MATCH(NEW_COMMAND))
    {
      // Send an INIT command to all subordinates.
      sendCommand(&heaterInitMsg, heater_sub_num);
      sendCommand(&levelInitMsg, level_sub_num);

      stateNext(S1);
      // Reinitialize variables here.

    }
  // Wait for all subordinates to report done.
  else if(STATE_MATCH(S1,
      heater_status->status == RCS_DONE &&
      level_status->status == RCS_DONE &&
      1))
    {
      status = RCS_DONE;
      stateNext(S2);
    }
  else if(STATE_MATCH(S2))
    {
      // Idle State
    }
}
```

Note that to send a command, first we must define an object of the command message class that we want to send. For example, in the code above the HEATER_INIT and LEVEL_INIT messages are defined. After that we assign any needed values to the data variables of these message classes. Note that the `INIT` messages of both the level and heater modules do not have any data fields. Therefore, assignment is not performed in the code in Example 2.11. After the values of the variables of the message are assigned, it can be sent to a particular subordinate using the **`sendCommand`** function, which is a member function of the `NML_MODULE` class. The subordinate to which the command is sent is determined from the unique subordinate number (**`heater_sub_num`** and **`level_sub_num`** in the code above) provided to the **`sendCommand`** function. Recall that these numbers were assigned in the constructor of the module during the initialization of the NML communication channels to the module's subordinates (refer to Example 2.7).

2.5.4 Main Program

In this section we discuss the main program, specifically for the tank application. To do this, we explain the code and overall operation of the program.

The Code

Recall that in the RCS design tool we assigned all the modules of the process control experiment to the same execution loop with a cycle period of 1.0 second. Therefore, all of them run in a single executable with the same "main" function. The main function creates a cyclic looping structure that activates the modules once every sampling period. Example 2.12, which is the code of `src/main/tankmain.cc` and is created by the design tool (with a few modifications after that), shows the main function for the tank RCS controller.

Example 2.12: Main function of the controller

```
/*
  tankmain.cc --- This file provides the C++ main function which
  creates and runs the following control modules:

  HEATER_MODULE, LEVEL_MODULE, SUPERV_MODULE
*/

// Include Files
#include <stdlib.h>// exit()
#include <signal.h>// SIGINT, signal()
#include "rcs.hh"  // Common RCS definitions
#include "nml_mod.hh"  // NML_MODULE definitions
#include "heater.hh" // definition of HEATER_MODULE
#include "level.hh" // definition of LEVEL_MODULE
#include "superv.hh" // definition of SUPERV_MODULE
#include "d20class.h"   // DAS20 interface functions

// flag signifying main loop is to terminate
int tank_done = 0;

//signal handler for ^C
extern "C" void tank_quit(int sig);
```

```
void tank_quit(int sig)
{
  tank_done = 1;
}

// DAS20 interface
DAS20Class das20;

// main loop, running 3 controller(s)
int main(int argc, char **argv)
{
  set_rcs_print_destination(RCS_PRINT_TO_STDOUT);

  RCS_TIMER *timer = new RCS_TIMER(1.0);
  HEATER_MODULE *heater = new HEATER_MODULE();
  LEVEL_MODULE *level = new LEVEL_MODULE();
  SUPERV_MODULE *superv = new SUPERV_MODULE();

  // Reset the data Acquisition card to a known safe state
  das20.das20_master_reset();

  rcs_print("\n\nPlease turn on the power switch now. \n");
  rcs_print("Press any key to continue. \n\n\n");
  getchar();

  // set the SIGINT handler
  signal(SIGINT, tank_quit);

  // enter main loop
  while(!tank_done)
    {
      heater->controller();
      level->controller();
      superv->controller();

      timer->wait();
    }

  // Delete Modules
  delete heater;
  delete level;
  delete superv;

  // Delete Timer
  delete timer;

  // Reset the data Acquisition card to a known safe state
  das20.das20_master_reset();
}
```

Analyzing the code, you can see that it includes all the necessary header files and defines all the needed objects. Moreover, a signal handler function called `tank_quit` is created to serve to quit the application if `Ctrl+C` keys are pressed simultaneously. Then the code defines all the objects needed by the application, such as the object for controlling the DAS20 data aquisition card, the NML module objects, and a timer for the loop. The operation starts by resetting the data aqcuisition card and enters the main loop formed by the `while` loop statement that continues until we press `Ctrl+C` that sets the variable `tank_done` to 1, leading to the exit of the loop. On exit of the main function all the objects are deleted and the data aqcuisition card is reset.

Overall Operation

The best way to describe the overall function of the RCS application is to step through the program operation and describe what each function is doing. This helps to obtain a general understanding of RCS. Figure 2.17 illustrates the operation of the supervisor module.

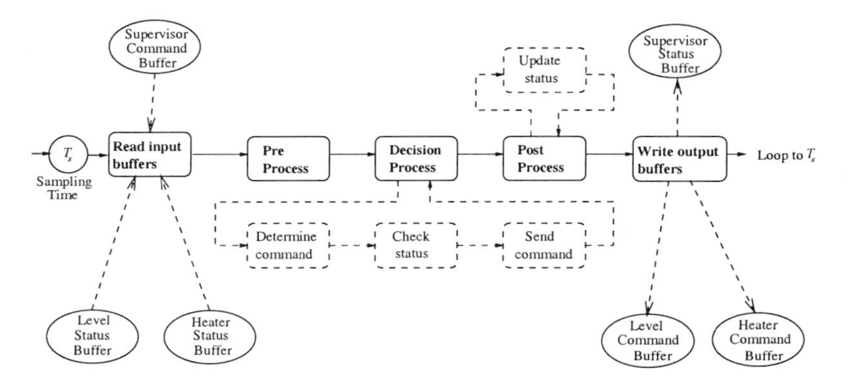

Figure 2.17: Overall operation of the supervisor module in the process control experiment.

Starting with the main program, the operation proceeds as follows:

1. First, an `NML_MODULE` object is created for each module. Therefore, since each module contains the code for controlling its part of the plant, the link between the plant and the processes is also established. Moreover, each module establishes the needed communication channels with its superior and subordinates. The communication buffers are created by the NML server since in the NML configuration file we defined the NML server as master for them. Therefore, the NML server must be run before this main program is run. (The NML server code is discussed in Section 2.5.5.) Moreover, at the start of the code we reset the data acquisition board, so that all actuators are off and the sensors are reset to start the operation. This allows us to start the program from a known and safe state.

2. Within the main control loop we establish the real-time environment with a cycle time of 1.0 second using the `RCS_TIMER` class provided by the RCS library. (The maximum resolution of the RCS timer depends on the system clock on the current platform.) The function `wait()` is a member

function of the RCS_TIMER class and puts the process to sleep until the end of the cycle time.

3. The function controller() is the main link to RCS control algorithms. It is a member function of the NML_MODULE class. For each module it needs to be called each cycle time and in simplified form accomplishes three important tasks:

   - It determines if a new message has been written to the command buffer of the module (i.e., if a new command has been received), and, if so, records the identification number of the message contained in the variable commandInData. Then it reads the status of the subordinates (if any) of that module.
   - It then calls the functions PRE_PROCESS(), DECISION_PROCESS(), and POST_PROCESS() in this order. The function PRE_PROCESS() allows the user to do any manipulation of data before control takes place. The DECISION_PROCESS() function calls the appropriate command function depending on the type of received command message. For example, if the command LEVEL_SET_REF is received in the level module, then the SET_REF member function of that module is called to execute the control algorithm for that command. Note that while executing the algorithm for a particular task the module may send commands to its subordinates. The POST_PROCESS() function allows the developer to do any necessary postmanipulation of the data. Here, we utilize this function to output the control value to the plant and set the status variables equal to the current measured plant variables so that accurate data is sent to the diagnostics tool, and to save them if necessary.
   - It updates and writes the status of the module to the status buffer, allowing the superior and the diagnostics to obtain updated information. Moreover, it also writes the new command messages to the command buffers of the subordinates.

4. Wait for the next sampling time and repeat.

In this application, we have all the processes running in a single executable with a sampling time of 1.0 second. However, if we wanted, we could run them as separate processes and different sampling rates could be used for each module. Note that in the process control case, the temperature module could be operated at a slower rate, since the dynamics of heating and cooling are much slower than those of adding or subtracting fluid from the tank. In this case, for example, we could set the sampling time of the level module to 1 second and that of the heater module to 10 seconds. The supervisor module should run frequently enough to calculate diagnostics of the lower modules (e.g., for motinoring or fault detection), although it will generally run slower than its subordinates.

2.5.5 NML Servers for the Buffers

In the sections above we explained the basic components of the RCS tank application. The tank experiment is a very basic example which does not involve a lot of modules distributed on many computers. We have one computer running the modules and the NML servers, and one computer for the diagnostics. In general, an RCS application can be more complicated; however, basic components of the code remain the same. In this section we provide an example code for the NML servers of the level module. Recall from Section 2.4 that we assigned all the buffers to the same process (executable), called `tanksvr`, that will run all the servers. Example 2.13, which is an excerpt from the file `src/main/tanksvr.cc` and is generated by the RCS design tool, shows part of the code of the main routine for this executable. We include only the code for starting the servers for the command and status buffers of the level module, `level_cmd` and `level_stat`, and the common `errlog` buffer. The code for the other parts of the program is similar.

Example 2.13: NML server for the heater module

```
/* tanksvr.cc --- This C++ file provides a main routin to start an NML server
   for this application. It connects to all of the NML channels used by this
   application. If they are all valid it will call run_nml_servers(), otherwise
   it will exit immediately.
   */
// Include Files
#include <stdlib.h> // exit()
#include "rcs.hh"  // Common RCS definitions
#include "leveln.hh"  // levelFormat

// NML Channel Pointers
static RCS_CMD_CHANNEL *level_cmd = NULL;
static RCS_STAT_CHANNEL *level_stat = NULL;
static NML *errlog = NULL;

static int InitNML()
{ // level
  level_cmd = new RCS_CMD_CHANNEL(levelFormat, "level_cmd",
  "tanksvr", "tank.nml");
  if(NULL == level_cmd)
    return -1;
  if(!level_cmd->valid())
    return -1;

  level_stat = new RCS_STAT_CHANNEL(levelFormat, "level_sts",
    "tanksvr", "tank.nml");
  if(NULL == level_stat)
    return -1;
  if(!level_stat->valid())
    return -1;

  errlog = new NML(nmlErrorFormat, "errlog", "tanksvr", "tank.nml");
  if(NULL == errlog)
    return -1;
  if(!errlog->valid())
```

```
        return -1;

    return 0;
}

static void DeleteNML()
{
    // level
    if(NULL != level_cmd)
      {
        delete level_cmd;
        level_cmd = NULL;
      }

    if(NULL != level_stat)
      {
        delete level_stat;
        level_stat = NULL;
      }

    if(NULL != errlog)
      {
        delete errlog;
        errlog = NULL;
      }
}

// Main
int main(int argc, char **argv)
{
    set_rcs_print_destination(RCS_PRINT_TO_STDOUT);

    if(InitNML() < 0)
      {
        DeleteNML();
        rcs_exit(-1);
      }

    run_nml_servers();
}
```

One can easily see that the program above tries to establish the NML communication channels and if successful, calls the function `run_nml_servers()`, which runs the servers for all the buffers and does not exit, or if unsuccessful, the program deletes all the communication channels and exits with error code of -1.

Recall that the master for a given buffer is responsible for creating this buffer and that we defined the process `tanksvr` (i.e., the program above) as the master for all the buffers in the NML configuration file. This also implies that this program should be run before running any other part of the application. Furthermore, since this program also reads the buffer and procedure names from the configuration file, the names of the processes and the buffers passed to the NML constructors should be taken from the configuration file. If you generate this file using the design tool, then you do not have to worry about the names to be passed, since the design tool will generate all the parts of the code in a consistent manner.

2.6 User Interface via the RCS Diagnostics Tool

In the preceding sections we completed the design and implementation of the RCS controller for the process control experiment. Now it is a time to execute and test proper operation of the system. Recall that we want a user interface from a remote host to all the running modules in this application. The RCS diagnostics tool is developed to serve this purpose. The RCS diagnostics tool is a Java-based graphical program that can be viewed using Java-compatible Web browsers or run as an applet or stand-alone application. Through this program the user can connect to the buffers in a running application, view the status of the modules, and send commands to them.

To run the RCS diagnostics tool as a stand-alone application, type

```
java -cp $CLASSPATH diagapplet.diagapplet
```

on the command line. Above, the environmental variable `CLASSPATH` should be set to the path for the RCS Java library and the Java Development Kit library.

2.6.1 Architecture File for the RCS Diagnostics Tool

The information needed by the diagnostics tool about the location of the RCS controller and to which modules to connect is provided by an architecture file. An architecture file is a text file which provides information about the structure of the controller, the buffer numbers, the communication ports, and the files in which the commands for the controller are defined. It can be generated by the RCS design tool, so it does not produce any extra programming overhead for the RCS designer. Example 2.14, which was generated initially as `src/tank.cfg` by the design tool and later modified and moved to `eepc99`, where the diagnostics program runs, shows the code for the architecture file for the process control experiment.

Example 2.14: Architecture file for the modules in the tank

```
heater{
        cmd_types="d:\\RCSproj\\tank\\src\\intf\\heatern.hh";
        stat_types="d:\\RCSproj\\tank\\src\\intf\\heatern.hh";
        SourceCodeDirectory="d:\\RCSproj\\tank\\src\\heater";
        nml_configuration_file="tank.nml";
        class_name="HEATER_MODULE";
        host=eepc100;
}

level{
        cmd_types="d:\\RCSproj\\tank\\src\\intf\\leveln.hh";
        stat_types="d:\\RCSproj\\tank\\src\\intf\\leveln.hh";
```

```
        SourceCodeDirectory="d:\\RCSproj\\tank\\src\\level";
        nml_configuration_file="tank.nml";
        class_name="LEVEL_MODULE";
        host=eepc100;
}

superv{
        child="heater";
        child="level";
        cmd_types="d:\\RCSproj\\tank\\src\\intf\\supervn.hh";
        stat_types="d:\\RCSproj\\tank\\src\\intf\\supervn.hh";
        SourceCodeDirectory="d:\\RCSproj\\tank\\src\\superv";
        nml_configuration_file="tank.nml";
        class_name="SUPERV_MODULE";
        host=eepc100;
}
```

Note that the `level` and `heater` processes are defined as children of the `superv` process. Moreover, some of the files of the application code are specified in the architecture file, such as the files where the command and status messages are defined. The diagnostics tool needs these files in order to know the command and status messages used in the application as well as their fields. We need to transfer the application files from the tank computer (`eepc100`), where we performed the design, to the Windows NT computer (`eepc99`), which we will use for diagnostics. The paths in the architecture file in Example 2.14 reflect the location of the code under the Windows NT computer.

Note also that the name of the NML configuration file is specified in the architecture file. From the configuration file the diagnostics tool reads the names of the buffers, their buffer numbers, and the communication port numbers in order to establish the needed communication channels.

2.6.2 Tank Diagnostics

In order to run the application properly, as discussed before, you need to run the NML servers first and then you can run the main program. The diagnostics program can be run on the computer that the application is running on or from any other host that has a network connection to the application. If it is run from a different host, then you need to transfer the files of the application to that host so that the diagnostics file can read them.

The RCS diagnostics tool has different options for choosing different views. Here, we will not describe all the options. We describe the tool in detail in Chapter 9. One option is to view the hierarchy of the controller with their current commands and state of operation. Figure 2.18 shows this for the tank controller. Note that on each module, the name, current command, and state of operation are specified. On this view, you can click on a particular module, and you will be shown the set of the available commands. You can select the command you want to send and send it to the module.

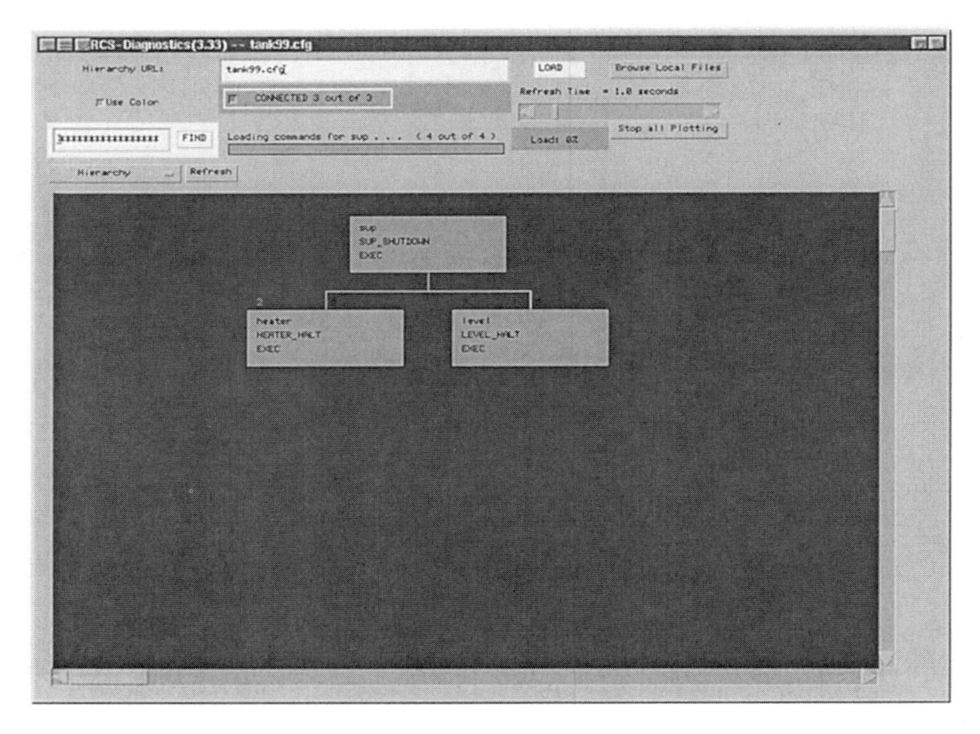

Figure 2.18: Hierarchy of the tank.

A better view for sending commands and viewing status is the "Details" view, which is shown in Figure 2.19 for the tank controller. From this view you can select the module from the list of the available "Modules" on the left side of the screen (where the level module is currently selected). Then from the list of "Available Commands" you can choose the command that you want to send (in the figure the command `LEVEL_SET_REF` is currently chosen), choose and modify its fields from the rightmost box (in the figure the field `desired_level` is currently selected), and send the command. Below that, you can view the fields of the current command and status messages of the module chosen. For example, in Figure 2.19 the `desired_level` of the `LEVEL_SET_REF` command, which is the current operation, is 3.0, whereas the `current_level` field of the status message of the module is 6.03.

If you want to plot the value of a particular field of a command or a status message of any of the modules, you can select that field and click on the "Plot this Command Variable" or "Plot this Status Variable" box, respectively. Figure 2.20 shows the plot of the `current_level` status variable during the operation of decreasing the level in the tank from 8 to 3. See Chapter 9 for more information on the RCS diagnostics tool.

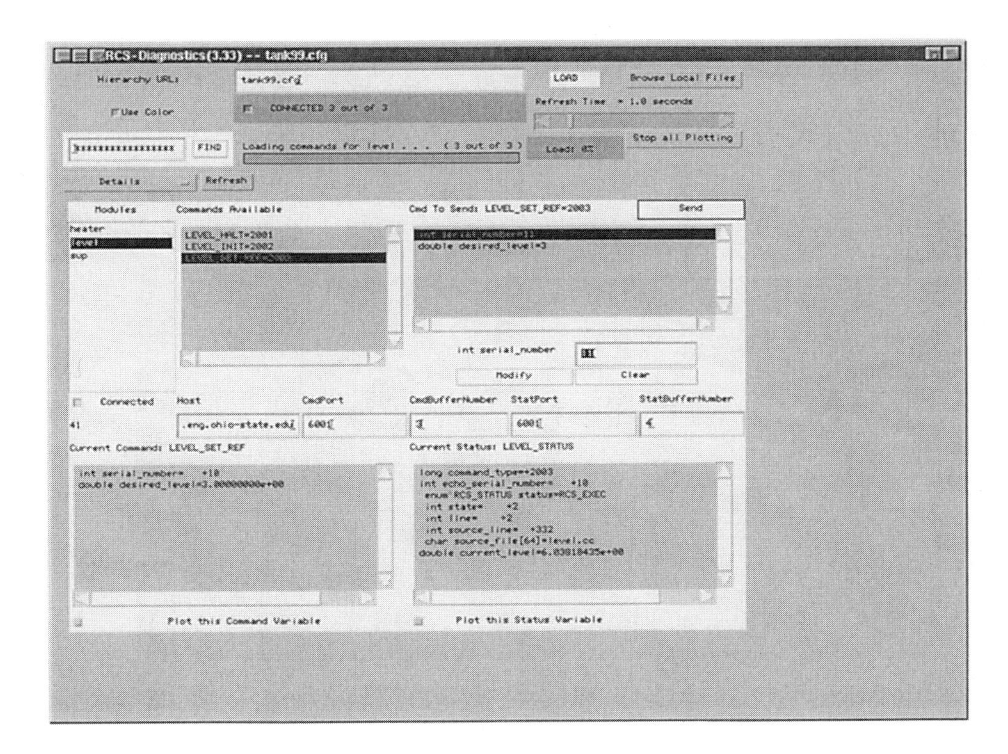

Figure 2.19: Details view of the tank.

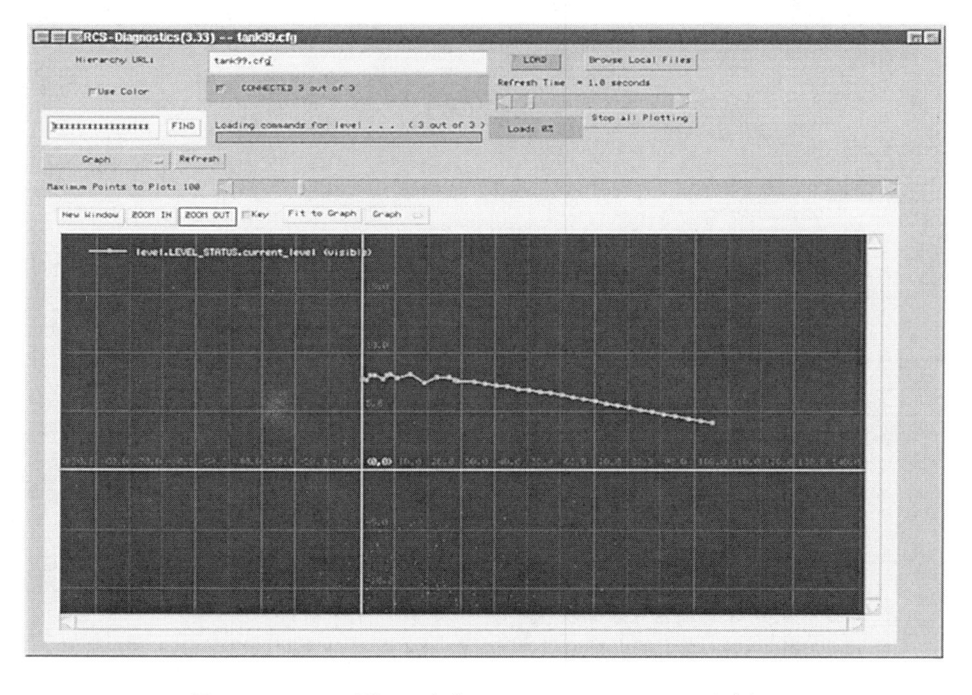

Figure 2.20: Plot of the `current_level` variable.

Chapter 3

Reference Model Architecture and RCS Applications

One of the advantages of the RCS software is that it facilitates the development and implementation of a variety of hierarchical control architectures. In this chapter we briefly introduce the *Reference Model Architecture* (RMA), which is a method for structuring and developing complex control systems. Moreover, we summarize some of the applications where RCS has been used to implement the RMA, and highlight the key features of RCS. It is important to emphasize, however, that although RCS was actually originally developed for implementations of the RMA, RCS is a very general design tool that can implement many different types of controller architectures (e.g., the *intelligent autonomous controller architecture* in [8, 56]). Using the ideas give on how to implement the RMA, it should be clear how to implement other controller architectures for other applications.

3.1 Reference Model Architecture

The RMA was developed at NIST by Albus and his colleagues via motivations from intelligent biological systems and the desire to achieve high levels of automation for a variety of practical problems (for a more detailed treatment of the RMA, see [7]). Often, in fact, the controllers that were implemented were developed using analogies with how humans performed tasks, and in some applications they replaced humans who performed tasks. In these cases, the control system that the RMA implements is sometimes called an *intelligent control system* since it is motivated by, or replaces, human intelligence. Generally, these intelligent control systems are designed to achieve highly autonomous behavior, and are able to reason and take actions that maximize the probability of achieving goals in unstructured and uncertain environments [7, 8].

In computer hardware and software or biological systems, intelligence requires at least the ability to sense the environment and take appropriate control actions. Higher levels of intelligence require the ability to recognize objects and events, represent the world internally, and generate plans for the future. Advanced forms of intelligent systems, ones that can achieve high levels of autonomy, have the ability to analyze, understand and learn, reason and plan successfully, adapt to changing conditions, and survive in complex and competitive environments. The level of intelligence achieved depends on the computational power of the computing machine (system processor or brain), the efficiency and complexity of algorithms for sensing, learning, recognition, planning, reasoning, and so on, and the quality of information and values stored in the system's memory.

Next, we outline the basic components of the RMA, where in each case you will see the connection to how the human performs certain tasks, some tasks that we perform almost every day (e.g., planning, learning, reasoning). For a detailed explanation of the connections between intelligent biological systems and the RMA structure and functions, see [7].

3.1.1 Basic Building Blocks for the RMA

The RMA is composed of a hierarchical interconnection of basic building blocks (e.g., blocks in the hierarchical controller, such as the components in Figure 1.1. We will refer to each component as a *processing node.* There can be multiple processing nodes at each level of the hierarchy, each processing node can have its own timing and may communicate with other nodes, but generally the upper levels of the hierarchy supervise the lower levels. To explain the operation of the RMA, in this section we explain how one processing node operates. Then, in the next section, we explain how an interconnection of these basic processing nodes can operate.

The structure of a processing node in the RMA is shown in Figure 3.1. The functions there are partitioned to four basic elements: *behavior generation* (BG), *world modeling* (WM), *sensory processing* (SP), and *value judgment* (VJ). The BG processes make the decisions and request the corresponding actions that are required to implement that decision. The WM processes hold the data and model of the environment. The SP processes gather information about the outside world. The VJ processes evaluate the costs and benefits of hypothesized actions. These are explained in more detail next.

Behavior Generation Processes

The BG process (Figure 3.1) executes tasks. A task is a piece of work to be done or an activity to be performed. For every BG process there is a set of tasks that this process can perform. Every task has a name, and the set of names of the tasks that the process can perform forms the task vocabulary for that BG process.

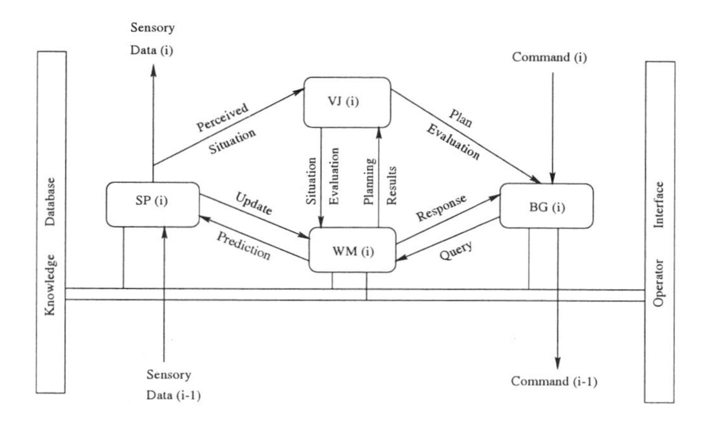

Figure 3.1: Basic building block of the RMA (figure from [8]).

BG processes are also known as *task decomposition* (TD) processes. Each BG process at each level consists of three subprocesses. These are *job assignment* (JA), *planning* (PL), and *execution* (EX):

- **JA subprocess**
  The JA subprocesses decompose a given input task command into N distinct jobs to be performed by N physically distinct subsystems, where N is the number of systems under control of this process. The JA process does not perform decomposition of the task in time. It decomposes it just in space. It is also responsible for resource allocation to the subsystems for their use in completing the task.

- **PL(j) subprocesses**
  The PL(j) subprocesses, $j = 1, 2, ..., N$, are associated with each of the N subsystems. Each of these subprocesses is responsible for decomposing the task assigned to its subsystem into a time sequence of planned subtasks to be performed. Each planner subprocess is also responsible for resolving conflicts and mutual constraints between hypothesized plans of the subprocesses.

- **EX(j) subprocesses**
  The EX(j) subprocesses, $j = 1, 2, ..., N$, are associated with each planner PL(j). Their job is to execute the tasks planned by the respective planner. When a subtask in a given plan is completed, the executors step to the next subtask in the current plan. When all the subtasks in a given plan are completed, the executors step to the first subtask in the next plan. If for some reason there is a failure in completion of a subtask, then the executor branches to a preplanned emergency subtask and its planner simultaneously begins emergency replanning to be substituted for the failed plan. The executors generate output commands which become input commands to the BG processes for the corresponding lower-level processes.

Mutual constraints may arise in plans involving concurrent actions by different subsystems. For instance, some task may need concurrent or cooperative action by several subsystems, or a start event for a subtask may depend on a goal event of another subtask. In this case, both planners and executors should be coordinated for successful and efficient completion of the job. When a command for a given task is received by the BG process, it searches its library of task frames to find a task name which matches the command name. When a matching task is found, it is activated using the parameters passed by the command, and as soon as the requirements of the task are met, the execution begins.

World Modeling Processes

In order to be able to reason and take appropriate actions in an uncertain environment, we need information about the environment. The world model is the representation of the external world contained and maintained inside the RMA processing node. It acts as a buffer between SP and BG processes. This enables the BG process to act immediately, even though the information is not immediately available by the SP processes. Moreover, it enables SP processes to do recursive estimation based on the old information stored in the WM processes. The world model is hierarchically structured and distributed such that there is a WM process with a *knowledge database* (KD) in each node at every layer of the RMA control hierarchy. The WM provides functions that act on the KD, whereas the KD is a hierarchically structured, distributed, and passive data store.

Knowledge Database (KD): The KD is the data store which includes both a priori information about the external and internal environment before any action begins and a posteriori knowledge which is gained from sensory data as the action proceeds. It represents information about the states of the world, events occurring, time, space, and so on.

The world model includes not only information about the outside world but also about the intelligent system itself. For example, it may include attribute values of the objects, the priorities of different goals, parameters or equations of the kinematic or dynamic models of the system, state variables describing the system parameters, the state of currently executing processes, and so on. If there is some discrepancy between the information in the KD and the data read from the sensors, then the KD is updated so that it is kept up to date. Information in the world model's knowledge database may be organized as state variables, system parameters, maps, or entity frames (a structure which stores information about a given entity in the system).

Jobs of the WM processes: Here is a summary of the actions performed by the WM processes:

- **Maintenance of the KD**
  WM processes keep the KD current and consistent. They update KD state estimates based on the correlation and difference between the world model predictions and sensory observations at every hierarchical level. They enter newly recognized entities, states, and events into the KD and delete the events and states that are determined not to exist by the SP processes. They also enter the estimates of reliability of KD data generated by VJ processes.

- **State prediction**
  Based on the information in the KD, world model processes estimate the sensory processing information expected. This enables the sensory processing to do correlation and predictive analysis of the data.

- **Data supply to other processes**
  The WM processes are responsible for supplying planners and executors of BG processes with information about states, events, and entities. The information about the current state of the world, supplied by the WM, is used by the executors for correct action and by the planners as a starting point of planning.

- **Simulation**
  In order to plan future actions, planners need to know the possible results of planned actions. It is a job of the WM processes to simulate the system for tentative plans and to supply the results to the VJ processes for evaluation. The BG-WM-VJ loop enables BG planners to select the best actions as the plan to be executed.

Sensory Processing Processes

The SP processes get information about entities, events, states, and relationships in the external world to keep the world model accurate and up to date. They maintain the correspondence between the internal world model and external world. The SP processes are not only responsible for obtaining data from the sensors, and filtering and integrating it, but also for interpreting this data. At a lower level this is implemented as feature extraction, pattern recognition, image understanding, and so on. At the upper level of the hierarchy more abstract representation and interpretation of data are implemented. Each SP process performs functions such as coordinate transformations and comparisons (between gathered data and stored information). Each SP process at each level uses the information from the lower levels to combine them for detection of objects of higher complexity.

Value Judgment Processes

In the subsections above it was mentioned that VJ processes compute the uncertainty in detecting events or objects. Other functions of VJ processes are to compute the costs, risks and benefits of plans and actions, the desirability, attractiveness, and repulsiveness of objects, and so on. VJ processes can evaluate events as trivial or important so that trivial events can be safely forgotten in order to free memory, and important events can be remembered for future reference. Moreover, the VJ processes can have judgment functions to indicate what to reward or punish, what should be treated as friend or enemy, and so on. It may assign different priorities to different goals such that objects defined as friends should be defended or assisted and enemies should be attacked or avoided.

The VJ state variables can be preassigned (or may be interactively assigned) by a human operator or may be computed on real time by the VJ processes themselves. To each entity or event, a value judgment state variable can be assigned so as to label it as expensive or cheap, risky or safe, good or bad, and so on. Plan selective processes may then select plans which are good, safe, and cheap. VJ processes are essential to the implementation of optimization processes in the RMA.

3.1.2 RMA Hierarchical Organization and Timing

In the RMA structure the processing nodes (i.e., building blocks as in Figure 3.1) are layered in a hierarchical graph or tree as shown at the top of Figure 3.2. Every node contains BG, WM, SP, and VJ processes. These processes are tightly interconnected to each other with a communication system. For example, in Figure 3.2 the process labeled B is connected to the process A, which is higher in the hierarchy, and processes C and D, which are lower in the hierarchy. At the lowest level of the hierarchy, the communications are often completed using common memory or message passing between processes on a single computer. At middle levels, communications are often implemented between multiple processors through a communication bus. At higher levels, communications can be implemented using gateways and local area networks. All nodes need not be present within each branch of the control tree; those that are not needed can be removed and can be thought of as ("unity gain pass through") degenerate computing nodes. The exact interconnection and branching in the control tree depends on the algorithm chosen for decomposing a particular task. Therefore, for the same application, different task decompositions may lead to different control tree structures. The task decomposition algorithm also affects the information that must be shared between different tasks.

Every hierarchical system may have a different number of levels in its structure; however, RMA suggests a seven-layered structure, as this has been found to be useful in several robotics applications. The functionality of each level can be derived from its characteristic timing or vice versa. The possible levels from top to bottom are *group level 7, group level 6, group, individual, elemental move, primitive*, and *servo*.

Levels in the RMA command hierarchy are defined by decomposition of the goals in time and space into levels of resolution. Moreover, the aggregation of sensory processing in time and space also contributes to the definition of the

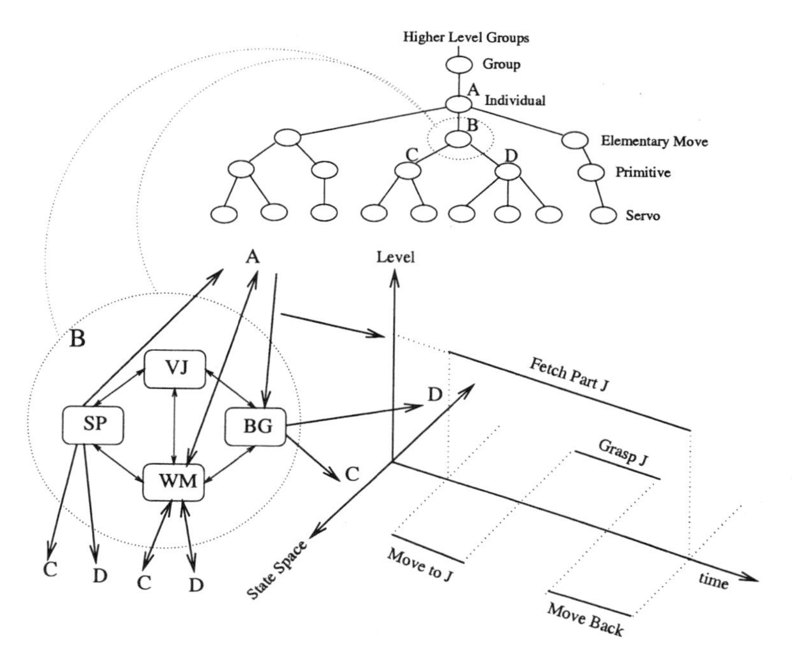

Figure 3.2: Organization and timing in the RMA.

levels in the control hierarchy. The resolution in time is implemented by loop bandwidth, sampling frequency, and state change intervals. The span in time is defined as the length of past traces and the length of planning horizon. The resolution in space is implemented by resolution of world maps and grouping of elements in subsystems. The span in space is defined in the range of world maps and span of control. At lower levels, the processes in the RMA structure become faster. That is, the sampling rate, the command update rate, the rate of subtask completion, and the rate of subevents increase in lower levels of hierarchy and decrease in upper levels of hierarchy. The amount of decrease and increase in every level is determined by the particular application.

Assume that process A sends a task request "Fetch Part J" to process B as shown in Figure 3.2. This task is decomposed by the BG processes of control node B to three different tasks, "Move to J," "Grasp J," and "Move Back," which are to be performed in sequence in time. The operations "Move to J" and "Move Back" are to be performed by node or subsystem C and the task "Grasp J" is to be performed by subsystem D as shown in the figure. This example shows the task decomposition in time and space. It also shows that the timing of the lower levels is faster than that of the higher levels. This is because the duration of the operations in the lower levels is smaller than the duration of tasks in the upper levels, which requires faster cycle rates. Next, more timing issues are discussed.

Figure 3.3 shows a timing diagram of task decomposition and sensory processing in the seven-layered RMA structure. The numbers in the figure illustrate the relative timing between the levels. Each application can have its specific timing based on the requirements. As you can see from the figure, the sampling rate, the command update rate, the rate of subtask completion, and the rate of subgoal events increase at the lower levels of the hierarchy and decrease at the upper levels of the hierarchy.

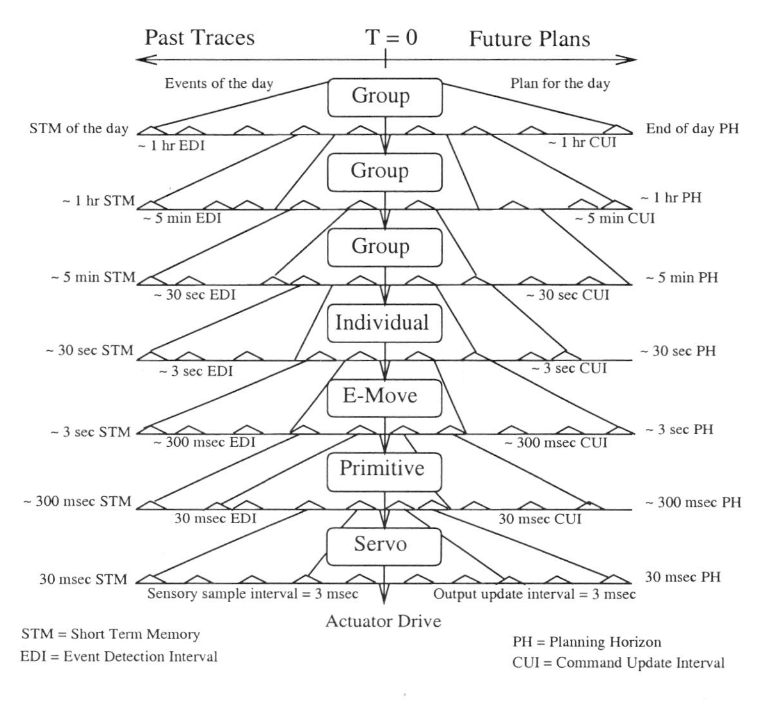

Figure 3.3: Timing diagram for activity flow in RMA (from [8]).

Planning is an important feature of the RMA, and there it proceeds simultaneously and continuously at all levels of the hierarchy. At each level, planners generate plans periodically, depending on the timing of this level. At each successive level, the first one or two steps of the current plan are further decomposed into subplans with higher resolution and less span in time and space (which is consistent with the timing of the level). This leads to a decrease of average period between task commands in each lower level.

Replanning is generally done on cyclic intervals. The interval of replanning can be the same as the average command update interval, or about 10 percent of the planning horizon. This makes it possible for the planner to generate plans as often as the commands are executed. Increasing the replanning interval may decrease the computational burden, but may also reduce the effectiveness of the system.

Execution of the tasks also goes on simultaneously and continuously at all levels. At each level, the executors periodically sample the feedback, compare it with the current plan, and send appropriate commands to correct the deviations between the plans and observations. Also, at each sampling period the data obtained via feedback is checked for emergency conditions. As soon as such conditions are detected, emergency replanning is started and preplanned emergency recovery routines are invoked immediately. This allows the system to respond to emergencies within a single cycle, and using preplanned routines first, creates time for the planners to replan the recovery plans.

3.1.3 Implementation of RMA Using the RCS Library

There are various ways to implement the RMA using the RCS library. You can define every control node in the RMA structure as an NML module or define the parts of a control node (i.e., BG, SP, WM and VJ) as different NML modules. After that, you should create communication channels between the NML modules according to the chosen implementation and the needs of the particular application. The hierarchical order can be created by defining some of the NML modules as subordinates of the others and giving access of the buffers of lower-level nodes to the upper-level ones. In particular, you can define two communication channels between a supervising node and its subordinate: one for sending commands by the supervising node to the subordinate and one for sending status information by the subordinate to the supervising node. CMS/NML will perform the communications between the modules using shared memory buffers. If the modules are on different computers, then the communications are done by using TCP or UDP protocols, and NML servers should be run for the processes that are remote to the buffer they need to access. If the computers are run by different operating systems and have different data representations, then the NML servers are responsible for encoding and decoding the data so that it can be interpreted.

The timing requirement for the modules in the RMA can be implemented using the timer functions that RCS makes available to the developer. Moreover, if there are mutual exclusion constraints in the application, they can be resolved using RCS semaphore utilities. It is a job of the application programmer to code the algorithms used within the control nodes of each layer. You can do the implementation using any algorithm since the RCS does not specify or limit the implementation details. Some applications may need a user interface through local computer or over a network. The RCS diagnostics tool, which is discussed in detail in Chapter 9, can be used for this purpose.

3.2 Summary of RMA/RCS Applications

During the past few decades the RCS library has been implemented and tested in many applications, including mining [32], manufacturing systems [6, 9, 10, 30, 35, 48], autonomous undersea and land vehicles [2, 31, 54], space station telerobotics [5], post office automation [3], and so on. With the development of the RCS software library, the design and implementation of similar applications becomes much easier and faster. The industrial applications may range from large-scale chemical plants, manufacturing lines, and so on, to small conventional control systems. In the universities it can be taught as a class and used in laboratory experiments [21]. Moreover, use of the RCS library is not limited to control system applications. It can be used in any application which needs interprocess communication. The process may or may not be hierarchical, and it may or may not be distributed over multiple computers.

In this section we summarize briefly some sample applications where the RCS library has been used to implement the RMA:

- **Intelligent autonomous vehicles**
  In recent years, the concept of an *automated highway system* (AHS) has gained considerable attention. This is because it brings numerous benefits to a society, including safer highways and higher highway throughput. Such a project requires design and implementation of intelligent vehicles. In this ongoing project NIST is developing an intelligent perception and vehicle navigation control system using RCS and RMA. RCS provides a systematic analysis, design, and implementation methodology for developing real-time sensor-based control systems. Functional task execution is viewed hierarchically with motor skill functions, such as steering and speed control performed at lower-levels and coordinated actions between vehicles performed at higher levels. The control system uses sensory information to guide the intelligent vehicle in the execution of complex tasks. Planning for task execution and for adaptation to changes in the environment are also part of the total hierarchy. Active and passive vision are the primary sensors for performing dynamic image perception analysis during navigation. Other sensors, including accelerometers, *inertial navigation systems* (INS) and *differential global positioning system* (DGPS) receivers, measure vehicle motion through the environment and provide precise localization of vehicles, targets, obstacles, and terrain features on a map database. The vehicle used as a testbed is shown in Figure 3.4.

Figure 3.4: Autonomous vehicle.

- **ROBOCRANE: an integration testbed for large-scale manufacturing and construction automation**
  Robocrane is a six-degrees-of-freedom work platform suspended by cables driven by winches under computer control, and it is shown in Figure 3.5. A PC-based *enhanced machine controller* (EMC) uses the RMA and is implemented via RCS. The controller has standard interfaces between functional modules for servo control, trajectory generation, constrained motion modes, discrete event logic, task sequencing, and an operator interface that incorporates the following elements: joystick control, a graphics programming environment, and a telepresence vision system with virtual reality displays that facilitate remote control by human operators. Construction of this system will lead to more efficient systems for users, greater market opportunities for equipment manufacturers, and lower costs due to greater competition between suppliers. Interface standards will mean easier, faster, and more robust integration of subsystems, less expensive customization of intelligent machine systems used in multiple applications, and lower costs for spare parts and system upgrades.

Figure 3.5: ROBOCRANE.

- **TETRA: a crane with enhanced control capabilities:**
  The goal in the *tetrahedral robotic apparatus* (TETRA), shown in Figure 3.6, project is to develop a new class of large-scale cargo manipulators, based on Stewart platform principles, which augment existing cranes or structures in order to improve operational safety, efficiency, and versatility of crane operations. TETRA uses the RMA and RCS for implementation. The principles of the Stewart platform parallel link manipulator are applied, using winches and cables as the links, to position crane cargo statically in all six degrees of freedom. Moreover, TETRA is used to study dynamic compensation of crane cargo. The concepts of active real-time control based on cable tensions and platform position to compensate for dynamic perturbations of the cargo are implemented. Adaptive control techniques incorporate a variety of sensors into the controller, such as tactile, proximity, and vision systems. This leads to stable crane operations and autonomous cargo-handling capabilities.

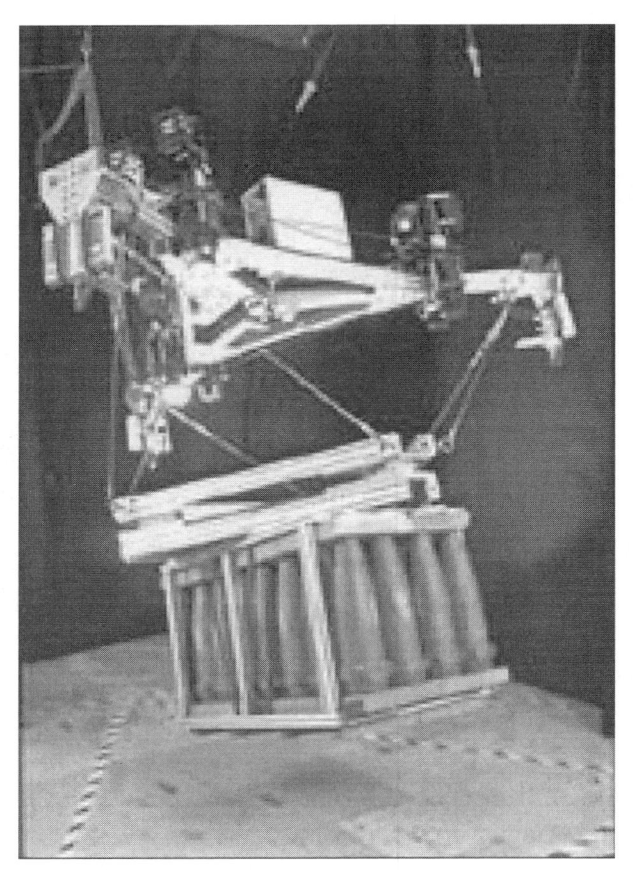

Figure 3.6: TETRA.

- **Next-Generation Inspection System (NGIS)**
  Inprocess and postprocess inspection with *coordinate measuring machines* (CMMs) often slows delivery of finished parts to the customers. The inspection of parts with complex geometry, where closely spaced inspection points are needed, is especially time consuming. The objective of the *next-generation inspection system*, shown in Figure 3.7, is to permit easier integration of advanced sensors into CMMs and to promote easier and more efficient integration of other controller components, such as motion control, obstacle avoidance, CAD software, and operator interface components. The testbed consists of a CMM, advanced sensors, and RCS. The advanced sensors include analog touch probes, a video camera, an analog capacitance probe, and a laser triangulation probe. The RCS controller permits real-time processing of sensor data for feedback control of the inspection probe. The controller also permits integration of a video camera for part feature recognition and for vision-guided motion control of the inspection probe. The controller will provide interfaces to CAD models and to the *dimensional measuring interface standard* to allow inspection that is driven by model data.

Figure 3.7: Next-generation inspection system.

3.3 Summary of RCS Library Features

Some of the most important features of RCS are the modularity of the approach and code, platform independency and portability of code, the easy user interface through the RCS diagnostics tool, and the fast, easy, and error-free code generation via the RCS design tool. In this section, which serves to close Part I, we discuss these advantages in a bit more detail.

- **Modularity for hierarchical distributed control systems**
  The RCS software library provides the tools to construct general controller architectures and hence RCS has been found to be useful for many applications. Any controller, from a conventional single-input, single-output proportional-integral-derivative (PID) controller to a complex hierarchical distributed controller can be implemented using RCS. Generally, for complex hierarchical controllers a task decomposition methodology is followed to construct the hierarchy, and hence to define the modules and their interconnections. Such an approach breaks the control problem down into smaller subproblems which are easier to analyze, develop, and implement. In other words, the user determines the layout and use of the RCS modules to minimize complexity and maximize performance, considering physical constraints and available computing resources. The resulting modularity leads to controllers that are generally easier to understand, modify, extend, and maintain.

- **Platform independency and portability of code**
  At the stage of implementation of a complex control system, in general, the designer faces several different practical real-life problems. Such problems typically arise from having a variety of incompatible computer hardware and software. The challenge for the designer is which platform to choose, how to make the code development generic and platform independent so that if there is a need for a change of a platform there is no need for redevelopment of the program, and how to communicate information between processes running on different and incompatible platforms. The power of the RCS software comes from the fact that it provides platform-independent code via its Communication Management System and Neutral Message Language. It allows different processes on different platforms to be able to "talk" to each other via shared memory buffers. This allows for a distributed and hierarchical control of an arbitrary number of subsystems by linking several modules across multiple backplanes or over a network. Moreover, each module in the hierarchy can be set to have a different cycle time without any complication or extra overhead in the code. This allows different subsystems of a large distributed system, with different timing requirements, to be easily connected in a single architecture.

  Since the code in RCS has a uniform structure it is possible to port and reuse previously developed code in different applications without making major modifications. Moreover, since the RCS library is platform independent, it allows you to port and reuse code even across different platforms or operating systems. The use of NML configuration files for specifying of the size of the buffers, mutual exclusion techniques, port numbers and communication protocols, and so on, allows the user to be able to change these without a need for re-compilation and re-linking of the entire application. These features bring several benefits to the RCS application developer. First of all, the programmer is freed from dealing with the time consuming and tedious interprocess communications which may also require knowledge of network programming. Moreover, portability of the code saves the programmer the effort of reprogramming the same algorithms for different platforms. Because of these features, the RCS library can be used not only for control system applications, but also for any application which needs two or more processes to share information. The processes could be located on single or multiple computers, with the same or different operating systems.

However, as the reader would agree, it is not possible to develop a code which is *completely* platform independent, due to the wide variety of available platforms and the fact that new ones become available at a regular pace. Therefore, there is a limit on the platform independency feature of the RCS library. Currently, it is available for most commercially used platforms. Appendix E shows the platforms on which the RCS library has been successfully tested.

- **User interface**
Using the RCS diagnostics tool it is very easy for a human operator to interface with the control algorithms at every level of the hierarchy. By choosing the different options available in the diagnostics tool we can view the fields of the status and command messages (i.e., the information passed between the modules), modify the command message fields and send them to the modules in the application, and monitor process data and plot it.

  Because of the platform independency and communication tools of the library, via the diagnostics tool the operator need not be on the workstation directly controlling the process, but can be connected to the application via the Internet virtually anywhere on the globe. Consider a chemical plant in which it is dangerous to have a technician in the field, but we still need the conditions in the chemical process, such as pH factors, to be checked and appropriate actions taken. With the use of the RCS diagnostics tool you can often do this directly from your office. By sending appropriate commands to the process, you can change reference set points or controller parameters, monitor process variables, or you can even switch from one control algorithm to another.

- **Automatic code generation tools**
Using the RCS design tool the programmer can automatically generate almost all of the application-independent code of a given application. Also,

it can be used to design or modify an entire RCS application structure, add or remove modules, set up or delete communication channels between modules, define messages to be passed between the modules, and so on. This can be done even for applications distributed over several computers with (possibly) different operating systems. The design tool makes it possible to generate the skeleton of the application together with most of the application-independent code in very short time. Moreover, it also generates the files used by the diagnostics tool. Therefore, there is no programming overhead for using the diagnostics tool. The only job left after creation of the skeleton of an application is implementation of the actual low-level control (or estimation) algorithms or state tables. Besides making the job of the programmer easier, the design tool contributes to the regularity of the code and makes the code developed less erroneous.

Finally, we would note that with the use of the *source code control system* or *revision control system* in developing code in RCS, multiple programmers can work on the same project simultaneously in their own workspace, without affecting the other programmers and the application code. Then, after testing and verifying the code that they developed, they can release their code to the application. This prevents a programmer from accidentally overwriting the code that a fellow programmer is developing.

Part II

RCS Handbook

Chapter 4

Design Theme Problem: Automated Highway System

In Part I of this book we introduced the reader briefly to the basic concepts in the RCS library, to some of the RCS library tools, and to RCS application programming. Moreover, we briefly described the RCS design methodology, the connections of the RCS library to some of the "intelligent" control system architectures available in the literature, and provided some RCS application examples. In Part II of the book we describe most of the features of the RCS library in detail. The reader may find some of the examples in this part of the book similar to the examples in Part I. This is not surprising since the code development in RCS is uniform and modular. This is one of the powers of the RCS library, because it makes the code easy to develop and understand. While to develop only a simple RCS application it is enough to read only Part I of this book, to better understand the RCS library utilities and to use them for more sophisticated applications, you will probably need to read Part II also. Following a strategy similar to that in Part I, in Part II we first introduce a design problem that we use for illustration purposes in the rest of the book. This problem is the automated highway system.

4.1 Introduction

Automated highway systems (AHSs) have received increasing attention due to the large amount of highway traffic congestion in major metropolitan areas [27, 39, 40, 45, 57]. AHS offer numerous potential benefits to society, including the reduction in traffic congestion and increased highway throughput and safety. The development of such a system is an extremely complex endeavor, due to the fact that the dynamics of a vehicle are complex, and since such a large amount of vehicle interaction possibilities exist (e.g., passing vehicles, multiple lane roads, traffic entering/exiting highways, etc.). Furthermore, each vehicle in an AHS must be quite autonomous. It should be able to plan and successfully complete the needed trips, learn and adapt to changing conditions (degradation of components, weather changes, etc.), and predict and prevent faults and bring the vehicle to a safe position in case of any. Communications is also of key importance in an AHS—the cars must be able to communicate with each other so that the coordination of activities can take place and no accidents occur. The individual vehicle's subsystems must also be able to intercommunicate to provide proper operation and to satisfy fault-tolerant and safety requirements.

Consider a case where a vehicle in an AHS has a problem. Specifically, let the car have a problem with the brakes. Obviously, a car with faulty brakes may cause accidents. In such a case the brake sensors should tell the braking subsystems that there is a problem, which should in turn communicate with the vehicle's intelligence center, letting the system know that there is a problem. The vehicle intelligence center must then decide what to do in this case. This may include further communications with the road-side controllers or platoon supervisor or leader (if these exist), indicating that the vehicle must pull off to the side of the road (in the case of extreme emergency). Or, the vehicle intelligence may decide that the problem is not that severe (maybe only one wheel is causing a problem) and decide that the car can finish its journey safely. This example shows the importance of intelligence and communication in an AHS. It shows that to obtain full autonomy, it is necessary to introduce some sort of intelligence into the algorithms. Clearly, a vehicle will encounter complex hurdles along its journey, including following a winding road, dealing with weather, vehicle faults, operations of the other vehicles, and so on. Such a system, for safety reasons, will also contain vast numbers of sensors to help it deal with and solve the problems introduced and situations encountered. Therefore, a hierarchical controller structure could be the best for dealing with all complex situations and the amount of sensory information that needs to flow in the system. As we discussed before, there are several hierarchical controller architectures [7, 8, 56] that are developed for controlling complex control systems such as an AHS.

An important feature of an AHS would be planning, since otherwise optimization of traffic throughput and minimization of accidents may not be possible. Planning horizons on different levels of the AHS hierarchy may be different. For example, while planning on the vehicle level may involve just a couple of possible future sequences of actions of the car (such as change lane–pass–change lane again), planning on a higher level may include trip planning from a given source to a required destination.

An AHS may need to adapt itself to the changing conditions. For example, if there are new highways opened for use and some of the old ones abandoned, temporary road closures and road constructions, and so on, should be reflected in the database of the system and the priorities should be updated accordingly for better utilization.

4.2 Possible Controller Designs

4.2.1 RCS Methodology–Based Design

In [4] the authors suggested a design for an automated highway system that was based on the RMA and RCS design methodology. There they suggest an RCS design for both vehicle control and highway control. In other words, in each vehicle there is an RCS control system and there is another RCS control system for the highway control. These systems interact and share information for fast and safe highway throughput.

Figure 4.1 shows the suggested RCS controller structure for an intelligent vehicle. The modules in the higher level of the controller hierarchy are re-

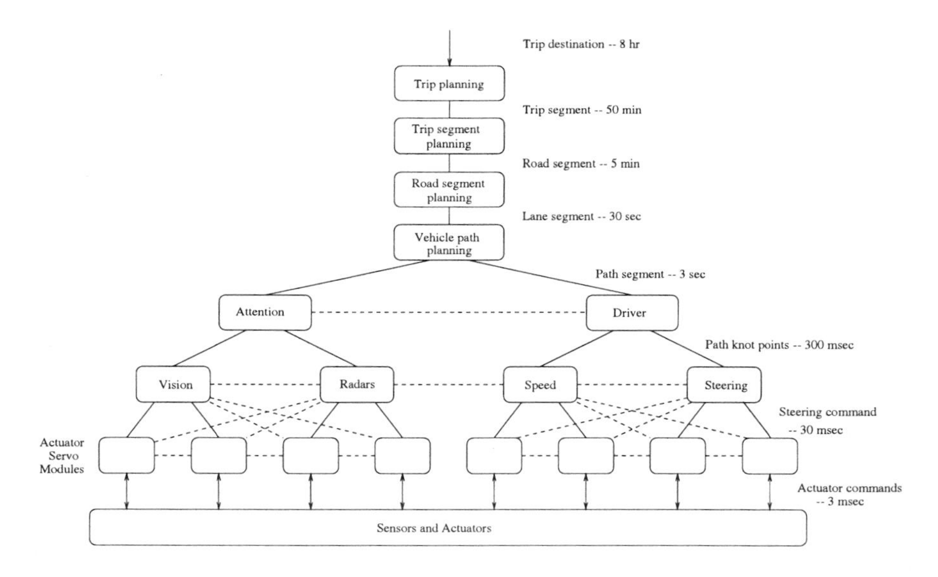

Figure 4.1: Controller hierarchy for an intelligent vehicle (from [4]).

sponsible for higher level tasks such as trip planning or road segment planning, whereas the modules on the lower levels are responsible from lower-level vehicle tasks such as speed control, steering control, radars and vision control, and the modules on the lowest level are responsible for the actuator outputs in the system such as brakes, throttle, vision system components, and so on. The solid lines in the figure represent communication channels between modules with a supervisor–subordinate relationship, whereas the dashed lines represent communication channels between modules that are not necessarily supervisor and subordinate. The figure also shows the relative timing horizons (requirements) of each level of the hierarchy. The numbers are for illustration purposes and the actual numbers may depend on the particular implementation. The relative timing shown in the figure is typical in hierarchical architectures. The resolution tends to decrease from lower modules to higher ones, and the range of the spatial maps tends to increase. Similarly, frequency domain measures of bandwidth tend to decrease from lower to higher modules. The low-level actuator and sensor modules will typically be operating at a much faster rate than the higher-level modules.

The trip planning module accepts (from the driver) a destination or a goal, and a possibly desired route, and generates a plan for the next 8 hours that consists of trip segments of about 50 minutes. The plans on this level could be generated by a computer system, may be supplied by a human operator, or may be generated by an interactive operation between the computer and the human operator. This module keeps maps out to about 1000 km with a resolution of about 2 km and contains information about cities, states, major highways, interchanges, service areas, and so on. Sensory processing elements on this level recognize these entities and verify the current position on the map.

The trip segment planning module accepts as input from the upper module a destination or a goal about 50 minutes into the future and generates a plan for these 50 minutes. The plan consists of a series of road segment durations of about 5 minutes. Maps in this module have a range of 100 km with a resolution of 200 m and contain information about the towns, secondary highways, road junctions, rest stops, motels, restaurants, and so on, along the planned trip segment.

The plans of the trip segment planning module are passed as inputs to the road segment planning module that divides the plan further into lane segments about 30 seconds in duration. This module has maps in the 10-km range and a 20-m resolution that contain information about roads and streets, groups of vehicles, intersections, and so on.

Given the inputs from the upper level, the vehicle path planning module generates a plan for a series of vehicle actions about 3 seconds in duration. Sensory processing elements on this level recognize cars, trucks, trees, pedestrians, curbs, signs, traffic lights, road markings, parking places, and so on. Moreover, they compute the attributes of the recognized elements such as distance and velocity.

Given the desired path or trajectory for the next 3 seconds the driver and attention modules coordinate to generate the plan for coordinated actions of the lower-level modules. This level keeps information on the entities such as road surface, barriers, obstacles, proximal surfaces of the nearby vehicles, and so on. Sensory processing elements on this level recognize these entities and calculate their attributes such as distance, orientation, vehicle clearance, and so on.

The modules on the second level are responsible for the control of speed and steering subsystems as well as the radar and vision subsystems. The sensory processing on this level recognizes the lane markings, road boundaries, and edges of obstacles. This level operates on a timing of about 30 milliseconds and generates inputs to the lowest-level modules.

The lowest-level modules are responsible for the sensors and actuators of the vehicle. The sensors in each vehicle are electromagnetic lane detectors, video and infrared cameras, radar receivers, and communication receivers. The actuators are steering, throttle, brake actuators, head lights, radar and infrared collision avoidance transmitters, communication actuators, and camera actuators.

Note that this architecture, in a way, emulates the way that humans drive a car. We plan which roads to take to get to a given destination. Then, given a particular plan, the basic tasks while driving are to control the vehicle speed and the steering angle to keep the vehicle on the road and to avoid accidents, and this structure has separate modules to perform each of these tasks. The attention module, on the other hand, is analogous to the vision system (the eyes) and hearing system (the ears) of the human driver keeping track of the position of the vehicle on the road and the situation around the vehicle . If at any point we change our plan for the path for the trip we modify our driving accordingly, as the controller in this architecture would do.

Certainly, this is not the only possible design for an intelligent vehicle. We encourage the reader to think of other possible designs.

The structure of the RCS controller for the highway control suggested in [4] is shown in Figure 4.2. It suggests that a large network of highways be organized

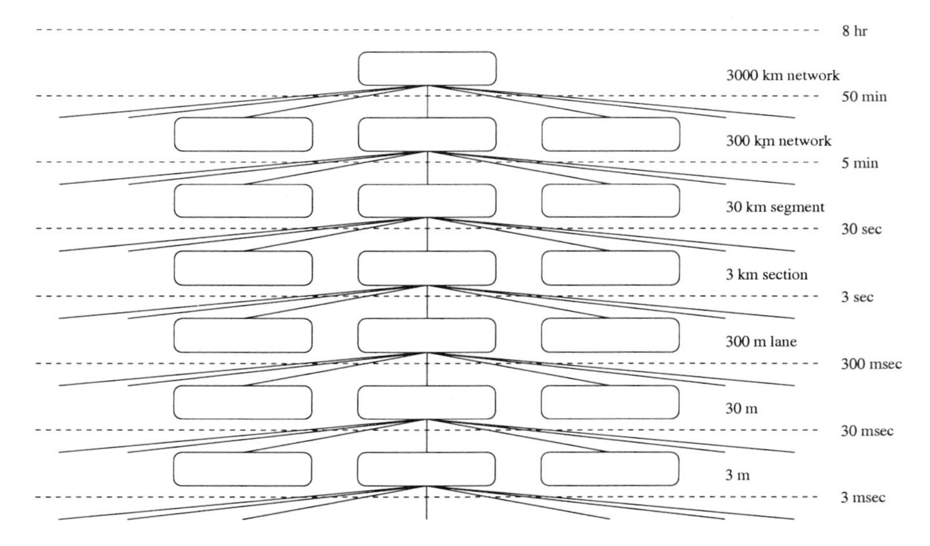

Figure 4.2: Controller hierarchy for an intelligent highway (from [4]).

into a hierarchy of smaller and smaller segments, sections, and subsections that are controlled from a control center on top of the hierarchy. As in Figure 4.1, the relative timing of each level of the hierarchy is also shown. The numbers on the figure are for illustration purposes and the actual numbers may depend on the particular highway for which AHS is designed. Note, however, the correspondence of the timing for the vehicle controller and the timing of the highway controller. For each section or subsection of the highway there is an RCS module that sets the current highway parameters for this section, such as maximum speed, minimum spacing, and so on. Such decisions are based on the current conditions in the section, such as weather conditions, the condition of the road, the (approximate) number of the vehicles in the section, and so on. These modules control the traffic lights, the programmable highway signs, radar transmitters, communication transmitters, and so on. They get information about the conditions of the highway section via various sensing devices, such as magnetic loop detectors, video and infrared cameras, radar receivers, communication receivers, and so on. For more information on the roadside controller, the reader may consult [4].

Note that the analysis and verification of the proper operation of the foregoing design for an AHS is not trivial. This is because there are a lot of different tasks that a vehicle can perform, such as "Move forward," "Move backward," "Accelerate," "Decelerate," "Move with constant speed," "Change lane," "Turn to right," "Turn to left," "Make U turn," and so on, and a variety of possible interactions between the vehicles, such as passing vehicles, changing lanes, vehicle suddenly moving in front of another vehicle, sudden braking by a vehicle in front, and so on. Therefore, to ensure safe operation there may be a need for "cooperation" between the vehicles. Below we describe another design that tries to simplify the problem by limiting the autonomy of the vehicles and the tasks that it can perform, and introducing a protocol that needs to be followed during each task.

4.2.2 Another Design Approach

The two RCS-based controllers described above are one possible implementation of an AHS. Other designs in the literature represent the road-side controllers and the vehicle controllers in a single controller structure. Such a controller design is described in [27], where the authors suggest a three-level hierarchical control structure, shown in Figure 4.3, that combines both vehicle and highway control. This design organizes the highway traffic into platoons (see Figure 4.4) for maximum highway throughput and safety. The architecture uses a variety of communications between the individual vehicles, platoons of vehicles, and the highway control centers. All these communications and vehicle operations are based on a predefined highway protocol.

In the hierarchy in Figure 4.3 the link layer controllers are on the roadside, whereas each vehicle has its own platoon and regulation layer controllers ("physical" in the figure refers to the actual vehicle dynamics). The reader may note the similarity of this design to the previous design, with the roadside RCS controller replaced with a link layer and the vehicle RCS controller replaced with a two-layer (platoon and regulation) controller. However, operation of the two systems is quite different since the second system organizes the traffic in platoons. Moreover, the connection between the the link layer and the vehicle controller in this design is somewhat tighter than the connection between the two RCS controllers in the previous design.

The link layer controllers are responsible for smooth flow of the traffic in each lane of the highway. They set paths for the vehicles, optimum speed and optimum platoon size. There is one link layer controller per highway section that is about 1 km long. Note that the physical layout of the roadside controllers is similar to the layout of the modules in the middle of the hierarchy in

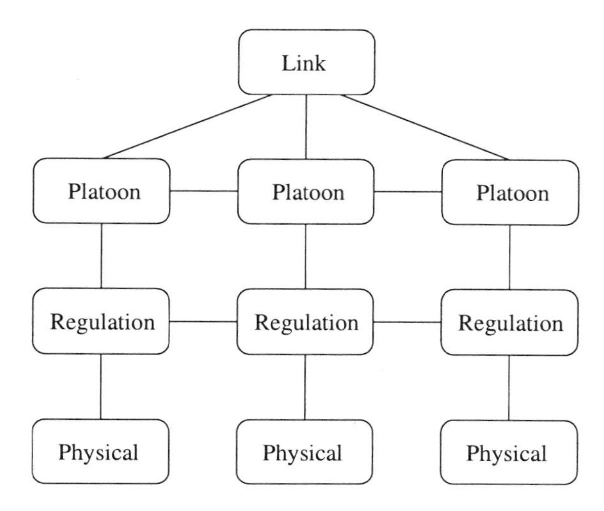

Figure 4.3: Controller hierarchy for an automated highway system (from [27]).

the earlier RCS methodology–based roadside controller. However, their job is quite different, since in the previous design the vehicle itself performs its path planning, whereas in this design it is the job of the roadside controllers.

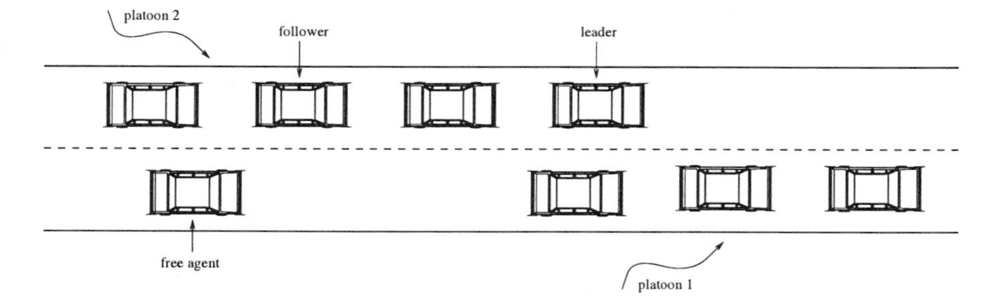

Figure 4.4: Platoon.

The platoon layer selects the maneuver to execute in order to follow the assigned path and coordinates the maneuver with the neighbors. The regulation layer implements the maneuver selected. There are three possible maneuvers that could be initiated by the platoon layer. These maneuvers are shown below.

- `merge` – In this maneuver two platoons join together to form one platoon. It is implemented by the first car, called the *leader*, of the second platoon (see Figure 4.4).

- `split` – In this maneuver one platoon splits to two platoons. It is implemented by one of the follower cars in the platoon. The vehicles that are not leaders are called *followers*, as shown in Figure 4.4.

- `change lane` – In this maneuver a vehicle changes lanes. It is implemented by a single-car platoon, called a *free agent* (see Figure 4.4). In other words, only free agents are allowed to change lanes.

The regulation layer can implement five feedback laws for vehicle control. Three of these are the algorithms for implementing the maneuvers above and the other two are

- `tracking law` – This is implemented by a platoon leader to keep the optimal platoon speed.

- `the follower law` – This is implemented by the follower vehicles to keep the optimal intervehicle spacing in the platoon.

Each maneuver is performed based on a predefined protocol. In other words, before each maneuver, the vehicles involved in the maneuver exchange messages to ensure the safety of the maneuver. In the merge operation the vehicles involved are the leaders of the two platoons that are merging together. In the split operation the involved vehicles are the leader of the platoon and the follower vehicle from where the the platoon will split. In the change lane operation, there can be two different situations. Assume that a vehicle (free agent) wants to change lanes. If there is a platoon or a free agent in the destination lane, then the vehicles involved are the vehicle that wants to change lane and the leader of the platoon that is in the destination lane. If there are no vehicles in the destination lane and there are vehicles in the lane adjacent to the destination lane, then the vehicles involved are the vehicle that wants to change lanes and the leader of the platoon that is in the lane adjacent to the destination lane. If there are no vehicles in both the destination lane and the lane next to it, then the vehicle can change lanes directly without need for communication or message exchanges with other vehicles.

The overall operation of the system will be as described in [27]. You will enter the highway from the rightmost lane (third lane) and join the traffic. After that you will initiate the automatic controller and will tell the desired destination to the onboard computer (the platoon layer) either by voice or keyboard entry. The onboard computer will communicate with the roadside controllers (the link layer) and the link layer will assign a path to the vehicle on either the second or the first lane. Then the driver will tell the computer to take control of the vehicle. When close to the destination exit from the highway, the computer will alert the driver to resume manual control.

Note that the design above limits the operations that a vehicle can perform to a very small set. This greatly simplifies the analysis, implementation, and verification of the overall system, although it reduces the autonomy of the vehicles. In other words, the vehicles in this design do not have much decision, planning, or adaptation capabilities. All these functions are done by the roadside controllers. Still it is a valid design, since the main objective in an AHS is not designing "intelligent" or fully autonomous vehicles, but it is the increased highway safety and throughput. Moreover, sometimes we may prefer a design whose operation is simple to verify over another designs.

The discussion above shows that there can be different designs for an AHS controller. Moreover, the controller could be designed using any procedure including the RCS methodology. Furthermore, based on the design objectives or the system requirements, we can come up with different controller structures using the same methodology. No matter what controller structure is used and no matter how it was designed, the RCS software library could be used to implement it. In other words, we can implement the controllers in both Figures 4.1 and 4.3 using the RCS library. In this book we do not show the reader how to design a controller for an AHS from scratch. Instead, we directly suggest a particular design and use this design in the illustrative examples in the rest of the book. This design will have some features of the designs in both Figures 4.1 and 4.3. In other words, the architecture of the controller will be similar to the one in Figure 4.1; however, it will implement the protocol of Figure 4.3 with only minor modifications. Moreover, the design in the next section is kept simple, since its objective is to serve as an illustrative example. The complete portions of the RCS library code to implement it can be found in subsequent chapters.

4.3 Controller Hierarchy for an Intelligent Vehicle

Consider the controller structure shown in Figure 4.5. It is the hierarchy of the controller residing in each intelligent vehicle. The vehicle *supervisor module* together with the *maneuver module* correspond to the platoon layer in Figure 4.3 (in [27] the platoon layer consists of four sublayers; here we implement them as two levels of RCS modules), the *drive module* corresponds to the regulation layer, and the *attention module* is for performing all the communications with the link layer, the platoon leader, and the other vehicles. The reader may argue that the design in Figure 4.5 is the simplified version of that in Figure 4.1. However, this is not true, because the duty of each module in the hierarchy is

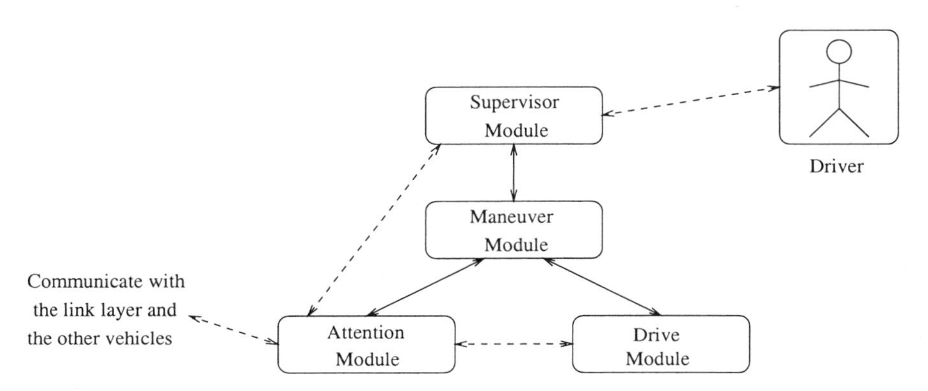

Figure 4.5: Controller hierarchy for an intelligent vehicle.

different from those in Figure 4.1. In other words, the overall organization of the highway traffic, the autonomy of the vehicles, and the task to be performed by the vehicles in the structure in Figure 4.5 follow those of Figure 4.3. The only difference from the controller in Figure 4.3 is that the communications of the vehicle with the link layer and the other vehicles are not done directly by the supervisor (platoon layer). Instead, they are performed through a special attention module. Therefore, one can think of the architecture in Figure 4.5 as an implementation of the controller in Figure 4.3 within the RCS framework.

Each vehicle has state variables that keep information about the vehicle state. Some of these variables are the following:

- `id` – the vehicle's identification number.
- `lane` – the current lane number.
- `section` – the current road segment or section number.
- `optsize` – the optimum platoon size.
- `optspeed` – the optimum platoon speed.
- `pltn_id` – the platoon identification number (this is the identification number of the platoon leader).
- `pos` – the vehicle's position in the platoon.
- `speed` – the current vehicle speed.
- `ownsize` – the size of the platoon the vehicle is currently in.
- `busy` – a flag to show whether or not engaged in an operation.

In addition to the variables above, the vehicle keeps information about neighboring vehicles and platoons. Below we describe the tasks of each module in a bit more detail.

4.3.1 Vehicle Supervisor Module

The duty of the supervisor module is to select the maneuvers to execute in order to follow the path assigned for the vehicle. It compares the vehicle's current lane and section with the assigned path and instructs the maneuver module to stay in the platoon or to leave the platoon and change lanes. Once again, lane changes can be done only if the vehicle is a free agent. Moreover, it compares the current platoon size, `ownsize`, with the optimum size, `optsize`, set by link layer, and the current speed to `optspeed`, and accordingly, asks the maneuver module to perform any necessary maneuver to maintain optimal values. Moreover, all the requests from the other vehicles are reported to the supervisor module by the lower modules. Based on these requests, the supervisor may decide to initiate a particular maneuver. The tasks or commands for this module are shown in Table 4.1.

Table 4.1: Tasks for the vehicle supervisor module.

Task	Description
INIT	Initializes the system: Initializes any needed variables and sends INIT to the maneuver module. Performed when the driver switches from manual to automatic operation.
HALT	Halts the operation: Saves any needed data and sends HALT to the maneuver module. Performed when the driver switches from automatic to manual operation.
AUTODRIVE	Normal operation: Runs as long as the automatic controller is on. Performed when the user tells the onboard computer to take control.

The overall operation of the system under this controller is as follows: When the driver initiates automatic control, the `INIT` task is executed. This task initializes the system variables, turns on any needed devices, takes the driver inputs, and performs any initial communications with the roadside controllers and the other vehicles. After that, the driver asks the computer to take control of the vehicle, which starts execution of the `AUTODRIVE` task, which continues as long as the automatic controller is on. After the driver switches from automatic to manual operation, the `HALT` task is executed to save the needed data, perform the final communications with the roadside controller and the other vehicles, and turn off any needed devices.

Figure 4.6 shows in state-transition-diagram form how the `AUTODRIVE` task is executed. We divided the diagram into three parts for brevity; however, the reader should see that the three parts correspond to a single diagram. Note

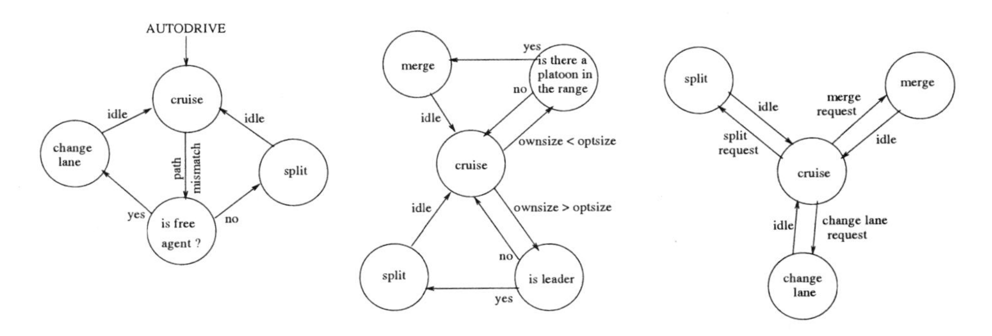

Figure 4.6: State transition diagram of the `AUTODRIVE` task.

that the signals for the request of any maneuver are initially received and processed by the attention module and passed to the vehicle supervisor through the auxiliary communication channel between them.

4.3.2 Maneuver Module

The maneuver module is responsible for safely performing the maneuvers that the supervisor module instructs and to coordinate the actions of the vehicle with the neighbors. The tasks or commands of this module are shown in Table 4.2.

Table 4.2: Tasks for the maneuver module.

Task	Description
INIT	Initializes any needed variables and sends INIT to the attention and drive modules.
HALT	Saves any needed data and sends HALT to the attention and drive modules.
CRUISE	Normal operation: The vehicle is not involved in any maneuver. It is either leader or follower in a platoon and implements the tracking or the follower control law.
MERGE	Join the preceding platoon: After proper communication with the platoon leader, accelerate until reaching the preceding platoon.
SPLIT	Split the platoon to two platoons: After proper communication with the platoon leader, decelerate until there is an appropriate spacing with the preceding vehicle.
CHANGE LANE	Move to an adjacent lane: After proper communication with the neighboring vehicles, implement the change lane law.

Before initiating any maneuver, the vehicles involved in the maneuver exchange messages. For example, consider the split platoon operation. It can be initiated either by a platoon leader or a follower. If the platoon size is greater than the `optsize`, then the platoon leader can send a split platoon request to a corresponding follower to match its platoon size to `optsize`. On the other hand, to change lanes a vehicle needs to be a free agent. Therefore, if a vehicle needs to leave the platoon, it may send a request to split the platoon to become a leader and then split the second platoon once more to become a free agent. The responding vehicle may grant the request, or if it is busy (involved in another maneuver) may deny it. Only one maneuver at a time is allowed. During any maneuver the supervisors of the vehicles involved in the maneuver enter the corresponding state and send the appropriate command to the maneuver module. For example, in the split operation above both vehicle supervisors enter the `split` state in Figure 4.6 (although in different parts of the figure) and send a `SPLIT` command to their maneuver modules.

Figure 4.7 shows the flow of the split platoon maneuver in state machine form for both the initiator and respondent vehicles. These diagrams are taken from [27] with little modification. Every maneuver can be represented similarly by state machines. For brevity, we do not show here the state machines of all the maneuvers. The interested reader may consult [27].

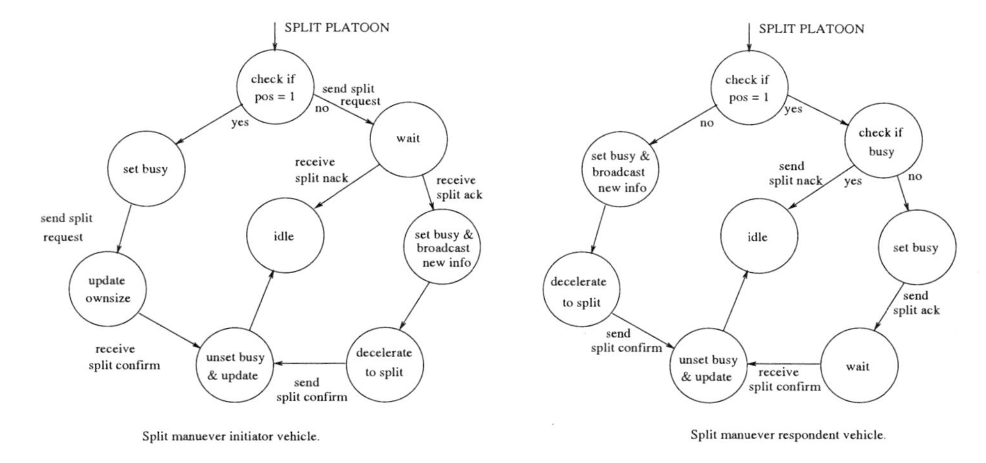

Figure 4.7: State transition diagrams of a SPLIT platoon maneuver for both the initiator and respondent vehicles (from [27]).

4.3.3 Drive Module

The drive module can implement five different feedback laws: accelerate, decelerate, change lane, follow, and track. Based on this, for this module we define the commands shown in Table 4.3.

Table 4.3: Commands for the drive module.

Task	Description
INIT	Initializes any needed variables and turns on any needed devices.
HALT	Saves any needed data and turns off any needed devices.
Accelerate	Implements the accelerate law. (Increases the speed of the vehicle.)
Decelerate	Implements the decelerate law. (Decreases the speed of the vehicle.)
Change lane	Implements the change lane law. (Moves to an adjacent lane.)
Follow	Implements the follower law. (Keeps a prespecified distance from the preceding vehicle.)
Track	Implements the tracking law. (Keeps the speed of the vehicle at `optspeed`.)

Figure 4.8 shows a simple diagram of how the DECELERATE task could be implemented. It simply suggests release of the throttle and use of the brakes only if there is a need. We encourage the reader to think of other possible implementations. You can develop similar diagrams for the other tasks of this module. Here we do not present other diagrams, for brevity.

Note that the drive module will generally need some low-level feedback signals, for example, position of the vehicle in lane change operations, the distance to the front vehicle in the vehicle follow operation, and so on. Therefore, we assume that the drive module has its own position sensory devices, such as radar

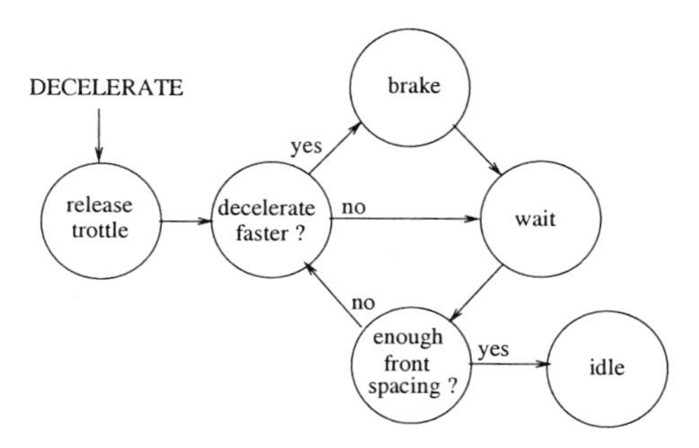

Figure 4.8: State transition diagram of the DECELERATE operation.

strip measurements, whose information is sufficient for implementing our longitudinal and lateral control algorithms or basically all five control laws. This allows the drive module to stay operational without having to read data from another module.

The drive module will also receive some sensory information from the attention module. This may include the speed and position information of nearby vehicles, including the vehicles on the side and rear. However, the longitudinal and lateral control algorithms will rely primarily on local onboard position sensors included within the drive module, whereas the information obtained from the attention module is for verification of the local information and enhancement of the algorithms.

4.3.4 Attention Module

The attention module is responsible for gathering the information about the world around the vehicle (e.g., surrounding vehicles and the road conditions) and performing all the communications or message exchanges of the vehicle with the link layer and the neighboring vehicles. (Note that this does not mean that it has control over all the sensors in the system. As mentioned before, the drive module has its own sensors, which are sufficient to implement its control laws.) It also maintains control of the vision and radar systems in the vehicle (if any) to keep the information about the surrounding world current. It reads and analyzes the intervehicle positions and spacings, dealing more with sensors that involve one vehicle's position with respect to another vehicle. Although it may add what seems to be redundant information to the overall system, this is necessary for safety reasons.

The communication receivers of the attention module are constantly polling for any incoming messages from the link layer or the other vehicles. Any information obtained is passed to its supervisor, the maneuver module. Moreover, some of the information is also passed directly to the vehicle supervisor and the drive module through auxiliary communication channels. The information passed to the drive module includes some sensory readings and information about the other vehicles. On the other hand, the same information and the incoming signals (messages) from the link layer and the other vehicles are passed to the vehicle supervisor. The sensor data gathered from the attention module is used by the supervisor primarily to determine which vehicles to communicate with before initiating a maneuver in order to ensure that actions that are about to be performed are safe (i.e., no collisions will occur).

We assume that the sensors that the vehicle is equipped with are capable of determining at least the following:

- The highway lanes and the lane the vehicle is on.
- The existence of a vehicle in front, if within the specified sensor range, its distance to our vehicle, and its relative speed.
- The existence of vehicles in the side lanes, their relative position, distance, and speed.

This information will be used for control purposes as well as to determine the possibility of merging to the platoon in front (if one exists) or to determine the vehicle to communicate with before a lane change operation.

If the vehicle needs to send a message to a neighboring vehicle or to a road-side controller, it does it through the attention module. The attention module is responsible for receiving and transmitting all the incoming and outgoing messages as well as all the encoding/decoding of data (if there is a need for any), filtering, and error checking. Table 4.4 shows the tasks that this module can perform.

Table 4.4: Tasks for the attention module.

Task	Description
INIT	Initializes any needed variables, turns on any needed devices, and registers the vehicle with the link layer.
HALT	Saves any needed data, notifies the link layer, and turns off any needed devices.
SEND REQ	Transmits a request signal for a maneuver.
SEND ACK	Transmits acknowledgment (approval) signal for a maneuver.
SEND NACK	Transmits not-acknowledge (disapproval) signal for a maneuver.
BROADCAST	Transmits broadcast information.

In addition to these tasks, each time a message is received by a receiver, a *receipt* signal for the message is transmitted back automatically if the message is recognized as a proper one, or a *retransmit* signal is transmitted back if the message is not recognized. We do not list that in the table of tasks since it is performed automatically and it would not be a command for this module.

Figure 4.9 illustrates the diagram of the `SEND REQ` task. In the state "prepare" the fields of the message, such as the `id` of the vehicle to be send to and the type of the request, are assigned, the message is encoded/decoded if needed, the transmitters are prepared, and so on. If for some reason the message fails

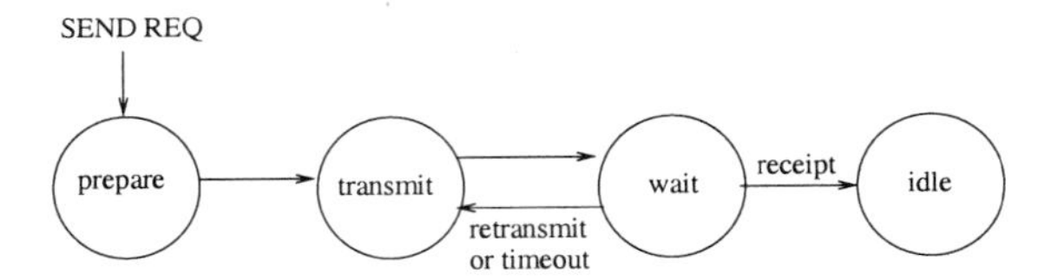

Figure 4.9: State transition diagram of the SEND REQ operation.

to go through in a proper form, the signal is retransmitted. The diagrams of the other tasks are similar. We leave them out for brevity.

4.4 Vehicle Dynamics

Design and implementation of a complete automated highway system is a long-term and expensive project. However, we can simulate the system to verify its proper operation. In order to emphasize the part of implementation using the RCS library rather than the development of the low-level control algorithms, assume simple point-mass vehicle dynamics of the form

$$\begin{aligned}\ddot{x} &= -\frac{\alpha_x}{m}\dot{x} + \frac{F}{m}\cos\theta \\ \ddot{y} &= -\frac{\alpha_y}{m}\dot{y} + \frac{F}{m}\sin\theta.\end{aligned}$$

In these equations, x is the longitudinal position and y is the lateral position of the vehicle, α_x and α_y represent the drag coefficients, m is the mass of the vehicle, F is the frictional force between the ground and tires (the drive force), and θ is the steering angle. For simulation purposes let $m = 1576$, $\alpha_x = 0.3$, and $\alpha_y = 0.1$ for all the vehicles. Note, however, that in an actual system the parameters of each vehicle would be different. For development of each of the five control laws of the drive module discussed earlier, we assume that:

1. Safety of the operation has been assured.

2. No faulty conditions are occurring.

3. Lateral and longitudinal velocities of the vehicle, and its lateral position, are known.

4. The distance to the preceding vehicle, as well as its relative velocity, are known.

5. The velocity of the vehicle during a lane change will remain (relatively) constant.

Note that a vehicle cannot change multiple lanes at once. In other words, the first lane change must be completed before requesting another lane change.

4.5 Low-Level Controller Development

Low-level control algorithms can be implemented using simple linear or nonlinear control methods, such as proportional-integral-derivative (PID), state feedback, sliding mode, adaptive controller, and so on. To illustrate, consider the lane change law and that we want to implement a state feedback controller for it. For simplicity we assume that the variables F and θ are our control inputs. In other words, there are no dynamics involved between the actual control inputs and the variables F and θ. Moreover, we assume that the lane change is performed during a road section that is straight or does not have curvature. Define $z = [\dot{x} - V_r, y - y_r, \dot{y}]^\top$ and $u = [F\cos\theta, F\sin\theta]^\top$, where V_r is the constant reference velocity (the `optspeed` set by the link layer) and y_r is the relative position of the center of the next lane that we want the vehicle to move in. With this we obtain

$$\dot{z} = \begin{bmatrix} -\alpha_x/m & 0 & 0 \\ 0 & 0 & 1 \\ 0 & 0 & -\alpha_y/m \end{bmatrix} z + \begin{bmatrix} 1/m & 0 \\ 0 & 0 \\ 0 & 1/m \end{bmatrix} u \tag{4.1}$$

since $\dot{V}_r = 0$ and $\dot{y}_r = 0$.

Let the desired closed-loop pole locations be $\{-1, -3, -2\}$ and choose $u = -Kz$, where K is a constant gain matrix. We choose the second pole the fastest because we want the vehicle to move to the next lane as fast as possible. The gain matrix

$$K = \begin{bmatrix} -3157.8 & 380.2 & 253.4 \\ 75.9 & -4727.9 & -6297.8 \end{bmatrix}$$

assigns the closed-loop system poles at the desired locations.

From the variable u we can get the actual control inputs F and θ using the equalities

$$F = \pm\sqrt{u_1^2 + u_2^2} \quad \text{and} \quad \theta = \tan^{-1}(u_2/u_1)$$

For driving forward we know that $F > 0$; therefore, we can uniquely determine F and θ from $u = [u_1, u_2]^\top$.

Based on the physical bounds on these system variables, we may know that $\theta \in [-\theta_{max}, \theta_{max}]$, where $\theta_{max} < 90^o$. Therefore, we may saturate the control input θ at $\pm\theta_{max}$. Similarly, if there is F_{max} such that $F \leq F_{max}$, we may need to saturate F also in order not to cause damage to the system. Below we chose $\theta_{max} = 1.2$ radians and $F_{max} = \infty$ in the simulations.

Figure 4.10 shows the response of the closed-loop system for two consecutive lane change operations. This may not be the best controller since it performs the lane change operation relatively fast and this may cause discomfort to the passengers. However, the purpose of this plot is to serve as an illustrative example.

The same controller can be used for a speed control operation, given that the current highway section on which the vehicle is traveling is straight. Figure 4.11 shows the response of the vehicle for an increase in the reference speed from 25 m/s to 35 m/s. Using similar techniques we can develop the controllers for the

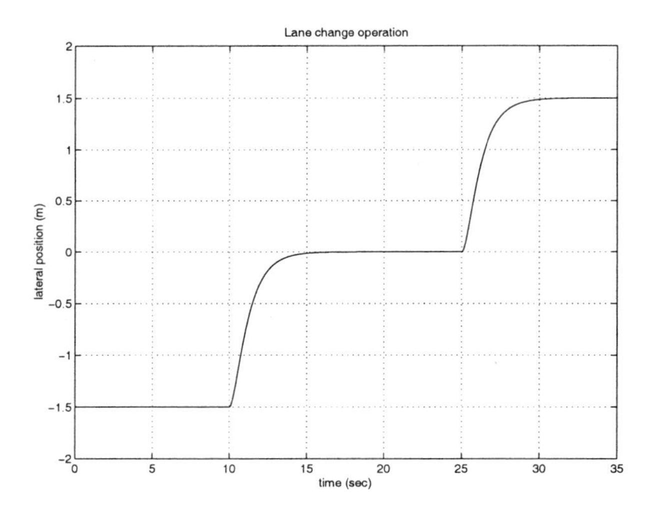

Figure 4.10: Simulation results for a lane change operation.

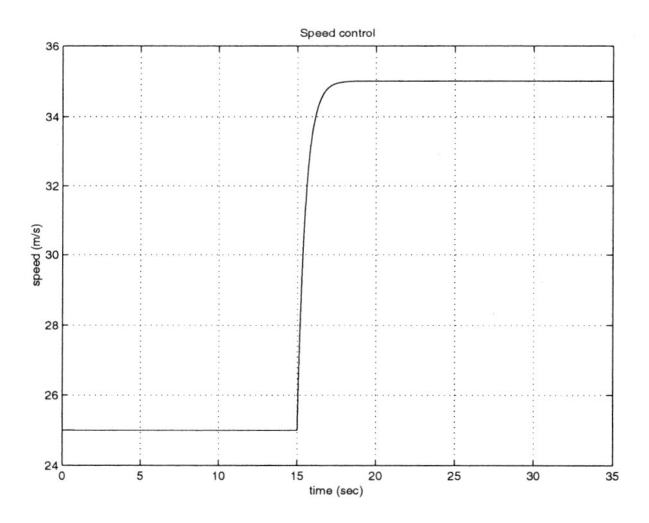

Figure 4.11: Simulation results for speed control.

remaining control laws in the system.

The controller design above just serves as an illustrative example for the reader. Moreover, the assumed vehicle model is a very simple one. There is a huge literature on modeling and control of an automated vehicle longitudinal and lateral dynamics. Some of the references to mention here include [17, 18] for modeling, [12, 36] for modeling and control, [1, 19, 20, 25, 26, 41, 42, 47, 55] for steering and lane keeping, [13, 14, 22, 24, 34] for lane changing, and [23, 49] for speed control and advanced cruise control.

In this chapter we presented a particular controller design for an intelligent vehicle in an automated highway system. The design followed the one presented in [27] with few modifications. The primary objective of this chapter was to introduce a design problem to be implemented using the RCS software library. Moreover, it also shows that the RCS library can be used on complex control systems such as AHS. Furthermore, the fact that we followed the design in [27] illustrates the it could be used to implement controllers that were not necessarily designed using the RCS design methodology.

In the following chapter we describe the RCS library in more detail and generally return to the intelligent vehicle described in this chapter when we need illustrative examples.

Chapter 5

Programming in NML

5.1 Getting Started

In this chapter we discuss the basic NML classes and functions that the RCS programmer needs to know. As we have seen before, RCS applications consist of hierarchically layered control modules which need to share information. For example, consider the simple RCS hierarchy of the intelligent vehicle in an AHS problem introduced earlier (see Figure 4.5). In this structure we have four control modules which are layered in three layers. In order to achieve correct operation, the modules in the hierarchy will need to be able to communicate with each other. At minimum, the supervisor module should be able to check the status of the maneuver module and also obtain information about the surrounding world in order to be able to carry its decisions. On the other hand, the maneuver module needs to know the status of the drive and attention modules so that it can perform the maneuvers safely. Moreover, each of these modules should be able to send task requests to its subordinates. For example, the supervisor module will send the requests for the required maneuvers to the maneuver module. Therefore, in practice we have to establish some means of communication between the modules or processes in an application. In the RCS library this can be done by setting up NML communication channels between the processes which need to communicate.

It is possible for any two modules in the RCS hierarchy to communicate with each other; however, more common communications are of a supervisor-subordinate type. For this reason, in general, we need to establish an NML command channel for transmitting commands from a supervisor to its subordinate, and an NML status channel for transmitting the status of the subordinate to its superior. Therefore, we need at least two NML channels between a module and its child or subordinate. The communications between the modules that are not in a supervisor–subordinate relationship are done through so-called auxiliary NML channels. Figure 5.1 shows the RCS library implementation of the controller for the intelligent vehicle (again, see Figure 4.5).

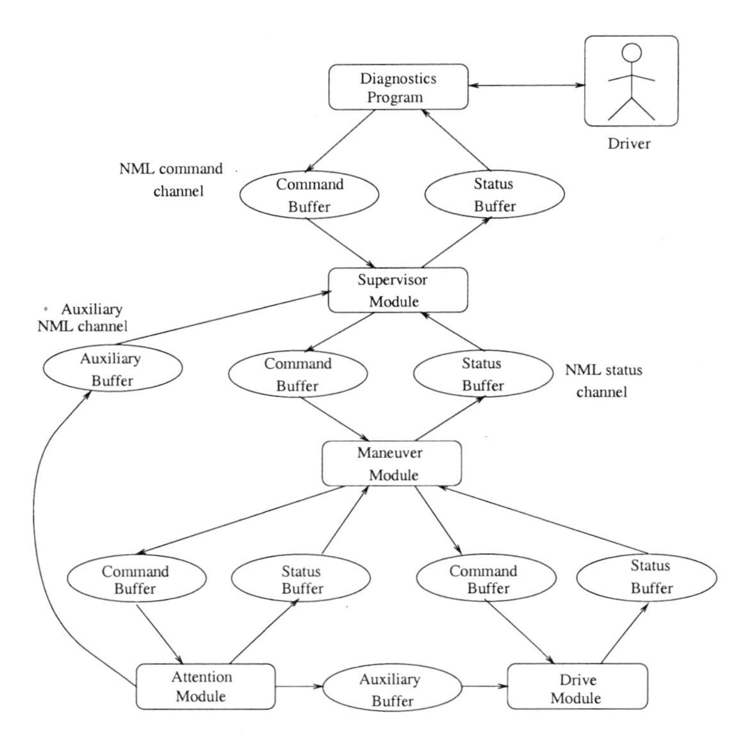

Figure 5.1: RCS implementation of the controller for an intelligent vehicle.

The communications in NML proceed by transmitting NML messages between the modules communicating with each other. NML has a predefined message class called `NMLmsg` which can be used by the programmer as a base class for deriving application-specific messages. This base class provides the basic data and function members which are present in every message; however, the RCS programmer decides on what other information should be included in the message.

There can be different types of messages transmitted between different modules in an RCS application. Two important types of messages are command and status messages. Command messages are those transmitted through NML command channel and carry information for a task request, and (possibly) parameters for the task such as desired or satisfactory conditions for initialization or successful completion of the task. They are, in general, passed from modules in the higher levels to those in the lower levels. For example, in the structure in Figure 5.1 the maneuver module can send a command "Change lane" or "Accelerate" to the drive module. The command "Accelerate" may carry information about the desired new speed, desired distance to the front vehicle, the duration of the operation, or some other data needed for successful initiation and completion of the task. Status messages are messages communicated through NML status channels and that carry information about the situation or status of the module, such as current coordinates or operating conditions of the module. In the vehicle structure in Figure 5.1, for example, the drive module may send the current speed, throttle, and steering angle to the maneuver module.

We have seen that CMS sets up access to fixed buffers of multiple read/write processes on the same processor, or over a network using a neutral code. NML provides a higher-level interface to CMS in an easier format to produce. Setting up an NML communication channel is, therefore, basically creating a shared memory buffer and setting the required access permissions to the processes which will communicate through this buffer. Then the modules share information through the buffer by one of them writing a message to the buffer and other(s) reading it from there. NML provides a uniform way for reading and writing messages to the shared memory. It provides a mechanism for handling multiple types of messages in the same buffer as well as simplifying the interface for encoding and decoding buffers in the neutral format and configuration mechanism. Although CMS is used for the communications between buffers and systems, we will be able to utilize the CMS tools without programming directly in CMS classes via the use of NML. In fact, the use of NML allows one to change, add, or delete buffers and processes without having to recompile the entire application.

Application programmers who use NML need to know several aspects of NML. You should know the basic predesigned C++ classes and functions that NML provides as well as how to use them for setting up communications between different control modules in the application. Detailed description of each of the classes and functions available in NML, together with examples of how to use them, are provided in the following sections.

5.2 Creating NML Message Classes

As we mentioned before, the RCS library contains a predefined message class, called `NMLmsg`, which can be used as a base for application-dependent messages. Now, we introduce the `NMLmsg` class and show how you can create your own message from it. After that, we will discuss the `RCS_CMD_MSG` and `RCS_STAT_MSG`, which are generic command and status messages.

5.2.1 `NMLmsg` Base Class

`NMLmsg` is a C++ class which is designed to supply the basic elements that each message should have. Its purpose is to ease the job of the programmer by making the message structure "uniform" and providing an easy way to create a communication vocabulary. It has several members, which we now list with a short description.

- `NMLmsg(NMLTYPE t, size_t s);`
  This is the constructor of the class. It should be supplied with a unique integer (`NMLTYPE`) identifier and size of message. In general, for every message we also define a unique identifier by adding _TYPE to the message

name and supply this to the constructor. (We discuss this in more detail later.) You can supply the `size_t` parameter by using the C function `sizeof` for the message itself.

- `NMLTYPE type;`
  This is the variable that stores the unique id of the message. This id number will be used to distinguish between different messages written to the same buffer.

- `long size;`
  This variable holds the size of the message. It is used so that the entire buffer is not copied unnecessarily.

- `void update(CMS *);`
  This is the update function. It is used to call CMS methods to encode and decode the message in neutral format. It is overloaded in each derived message in order to handle the new members of the derived class.

- `void clear();`
  When called, this function clears the message contents.

The message vocabulary that is used to create message classes is simply a set of C++ classes derived from the `NMLmsg` class. Having introduced the `NMLmsg` base class, we now discuss how you can create application-specific messages from it.

5.2.2 Deriving Messages from `NMLmsg`

The message class contains three main parts: a constructor which stores the unique identifier and size of the message, data variables, and an update function. The data variables placed in this class are the components of the message that are written to and copied out of the NML buffer during read and write cycles. As such, the NML message the user develops for a process in an application should contain the data that needs to be shared with other processes in the application (such as a diagnostics program or a supervisor module). These message classes could range in complexity from containing just a simple variable that is set to a certain value (say "1") if the control application is running, and another value ("0") if it is not running, to consisting of many elements that describe certain aspects of the plant that is being controlled (e.g., position, speed, power consumed). The application that reads the NML buffers determines what type of message is being read so that it can process the information correctly, hence the need for assigning the unique identifier. This identifier is simply a positive integer that determines what data is expected during a read or write. Therefore, each message class that is derived from `NMLmsg` must be assigned an identifier, by defining a type variable by adding `_TYPE` to the message name, that is unique within the memory buffer it is to access. Message identifiers need not be unique within the entire application, though it is recommended that they be.

Each message class must also contain an update function. This function calls CMS methods to convert the data members within the message class to a format that can be handled by the platform the program is running on. This function is contained in the RCS library and is equipped to handle most of the C-language basic data types. Note that we are required to create a new update function specifically for the new message type. All NML update functions (for each type of message) are identical except that the body of the function should call the CMS update for each member in the structure. Depending on the current mode of CMS, the update function either stores their arguments into a neutrally encoded buffer, or it decodes the information from the buffer and stores the output in the variables passed to the update functions. RCS uses format functions to identify which type of message is being handled and calls the appropriate update function, performing the desired conversion of data types. Example 5.1 shows the definition of a message class in a header file.

Example 5.1: Message class definition In this example an example message class, called `EXAMPLE_MSG`, is declared for illustration purposes.

```
/* nml_ex1.hh */
#ifndef NML_EX1_HH
#define NML_EX1_HH
#include "rcs.hh"

/* Give the new structure a unique id number */
#define EXAMPLE_MSG_TYPE 101
/* The id number must be unique within a CMS buffer, i.e. the number
   must be different than the id of any other type that might be
   written to a particular buffer. For simplicity it is recommended
   that the id number also be unique within an application. */

/* Define the new message structure */
struct EXAMPLE_MSG: public NMLmsg {
    /* The constructor needs to store the id number  */
    /* and the size of the new structure */
    /* by passing them as arguments to the base class constructor. */
    EXAMPLE_MSG(): NMLmsg(EXAMPLE_MSG_TYPE, sizeof(EXAMPLE_MSG)){};

    /* Each new type needs to overload the update function. */
    void update(CMS *cms);

    /* Data in this new message format. */
    float f;
    char c;
    int i;
};

/* Declare the NML Format function. */
int ex_format(NMLTYPE type, void *buf, CMS *cms);

#endif /* End of NML_EMC_HH */
```

Examining the header file, you can see that for the defined message called `EXAMPLE_MSG` there is a unique type definition `EXAMPLE_MSG_TYPE` through a

`#define` statement which is given the identifier 101. Furthermore, this class is derived from the `NMLmsg` class; this is necessary as the `NMLmsg` class provides the functions to write and read to the buffers. The reader who is not familiar with C++ should consult Appendix A for a brief introduction and [38, 53, 16] for detailed analysis on programming in C++ in order to have full understanding of the program. The message class `EXAMPLE_MSG` itself contains the data members `float f`, `char c`, and `int i` that are passed between the processes using this message as well as the constructor and an update function. The constructor needs to store the identification number and the size of the new structure. This is accomplished by passing them as arguments to the base class constructor. The update function calls CMS methods to convert the data members of the message to a type that can be handled by CMS. Example 5.2 shows the update function for the example message class `EXAMPLE_MSG` defined in Example 5.1.

Example 5.2: Update function for a message In this example the update function for the example message class `EXAMPLE_MSG` defined in Example 5.1 is presented.

```
#include "rcs.hh"
#include "nml_ex1.hh"

/* Create the update function */
void EXAMPLE_MSG::update(CMS *cms)
{
     cms->update(f);
     cms->update(c);
     cms->update(i);
}
```

In order for CMS to properly encode each type of message to allow NML servers to encode and decode data for remote processes, a format function exists to identify which type of message is currently being handled. This function consists of a simple switch structure that calls update functions for each specific message type. In some cases the user may only need a single message type, although in general there may be numerous types.

In Example 5.3 we present the corresponding C++ file with the code for the format function for the message class `EXAMPLE_MSG`.

Example 5.3: Format function for a message In this example the format function for the example message class `EXAMPLE_MSG` defined in Example 5.1 is presented.

```
#include "rcs.hh"
#include "nml_ex1.hh"

/* Add a new case to the format function. */
/* This is the format function that will be passed to the NML
   constructor. "type" is the id stored by the constructors of classes
   derived from NMLmsg in the type member. "buf" is a pointer to a
   block of contiguous memory where data to be updated is
```

```
   stored. "cms" is a pointer to the cms object that is used to access
   the globally accessible buffer. */

int ex_format(NMLTYPE type, void *buf, CMS *cms)
{
  switch(type)
    {
      /* . . Other Existing Message Types . .  */
    case EXAMPLE_MSG_TYPE: /* Add only this case. */
      ((EXAMPLE_MSG *)buf)->update(cms);
      break;
      /* . . Other Existing Message Types . . */

    default:
      rcs_print("ex_format: No Matching NML type.\n");
      return(-1);
    }
  return(0);
}
```

As we discussed before, the control modules in an RCS hierarchy, in general, receive and send at least two common types of messages. These are command and status messages. Command messages are job assignments that a module receives from its supervisor or sends to its subordinates. Status messages, on the other hand, are messages that contain information about the current working condition of the module sending the message. A module sends its status to its supervisors and receives the status of its subordinates. There can be other message types also; however, they are not as common as the command and status messages.

It is not difficult to imagine that the messages of the same type will have some similar or even same fields. For example, all status messages may have a field showing whether the module is still working on the job assigned, is done, is idle, or is in error. This suggests that definition of the base class for command and status messages containing common data and function fields would be useful. In fact, the RCS Library contains such classes, called `RCS_CMD_MSG` and `RCS_STAT_MSG`. We discuss them next.

5.2.3 Command and Status Messages

Command messages are those which are generally passed from higher-level modules to lower-level modules and request an action to be performed. Status messages are usually those which are passed from lower modules to higher modules, and contain information about the status of the module as well as (in some cases) crucial information necessary in higher-level decision making.

The `RCS_CMD_MSG` and `RCS_STAT_MSG` classes are both classes derived from the `NMLmsg` class. They are generic classes for use by the RCS application programmer. They include data members that are generally present in most of the applications, such as the command type that the module is working on, serial number, state of the module, and so on. The programmer can derive application-dependent command and status messages by deriving new classes from `RCS_CMD_MSG` and `RCS_STAT_MSG`. Another possibility is to derive the message classes directly from `NMLmsg` as discussed in the preceding section. However, if you use the `NML_MODULE` class (discussed in Chapter 6) in the application it is recommended to use `RCS_CMD_MSG` and `RCS_STAT_MSG` because they are part of the `NML_MODULE` base class.

Table 5.1 summarizes the common data members of the command and status message classes available in RCS, describing their functions. To read and write the message, the developer should call the NML basic functions for reading from and writing to a buffer, as well as numerous other read/write-related functions to, for example, determine if a new message has been written to a buffer since the last read, and so on. These read/write functions are discussed in detail in Section 5.4.

Table 5.1: Members of the command and status message classes.

Class Name	Members	Purpose
`RCS_CMD_MSG`	All members from `NMLmsg` plus	
	`RCS_CMD_MSG()`	(Constructor) Function that initializes variables and calls `NMLmsg()`.
	`serial_number`	Variable to store serial number of command.
`RCS_STAT_MSG`	All members from `NMLmsg` plus	
	`RCS_STAT_MSG()`	(Constructor) Function that initializes variables and calls `NMLmsg()`.
	`command_type`	Additional variables for RCS module use.
	`echo_serial_number`	
	`status`	
	`state`	
	`line`	
	`source_line`	
	`source_file`	

Note that in `RCS_STAT_MSG` there are fields such as `line`, `source_line`, and `source_file` which may seem redundant to the reader. These fields hold information about the source code file and are needed for diagnostics. In other words, these fields are used so that a human operator can see where within the program is the current execution. We discuss the RCS diagnostics tool in more detail in Chapter 9.

5.2.4 Creating Messages for an Intelligent Vehicle

Consider the RCS hierarchy of an intelligent vehicle in the AHS introduced before (refer to Figure 5.1). Possible command actions passed from the vehicle supervisor to the maneuver module would be MERGE, SPLIT, CHANGE_LANE, and so on, (see Table 4.2 for other possible messages). Similarly, the maneuver module will send commands such as ACCELERATE and DECELERATE to the drive module and SEND_REQ and SEND_ACK to the attention module (Tables 4.3 and 4.4 show the lists of the tasks for the drive and attention modules, respectively). Status messages for the drive and attention modules must contain data variables that must be communicated and shared with the upper-level modules for maximum performance. These may include current speed and position, intervehicle spacing, current lane of operation, any received requests, acknowledgments, and others.

These messages can be derived from the predefined RCS messages. In other words, the status messages are derived from the RCS_STAT_MSG class and command messages from the RCS_CMD_MSG. Example 5.4 shows the declaration of the DRIVE_FOLLOW command (this command corresponds to the task FOLLOW in Table 4.3; the DRIVE in front is added by the RCS design tool to make it easier to identify the module that performs the command) for the drive module of the intelligent vehicle.

Example 5.4: Sample command message for the drive module of an intelligent vehicle

```
// Command Message
#define DRIVE_FOLLOW_TYPE 4004

class DRIVE_FOLLOW : public RCS_CMD_MSG
{
public:
  //Constructor
  DRIVE_FOLLOW();

  // CMS Update Function
  void update(CMS *);

  // Place custom variables here.
  float optspeed;  // Desired vehicle speed
  float optspacing; // Reference inter-vehicle spacing
};
```

You can create the other command messages for the intelligent vehicle or any other RCS application in a similar fashion. Note that the variables or parameters in the command message show the desired speed and distance from the preceding vehicle. In other words, they specify the goal of this command. In some applications they may specify the initial conditions to be met for the operation to begin.

The declaration of the status message for the modules of an intelligent vehicle is similar. Example 5.5 shows the declaration of the status message of the drive module of the intelligent vehicle. The definition of the status messages of the other modules is similar.

Example 5.5: Sample status message for the drive module of an intelligent vehicle

```
// Status Message
#define DRIVE_STATUS_TYPE 4000

class DRIVE_STATUS : public RCS_STAT_MSG
{
public:
  // Normal Constructor
  DRIVE_STATUS();

  // Constructor used by derived classes
  DRIVE_STATUS(NMLTYPE t, size_t s) :  RCS_STAT_MSG(t,s) {};

  // CMS Update Function
  void update(CMS *);

  // Place custom variables here.
  float spacing; // Spacing between this car and car in front
  float x_vel, y_vel; // Longitudinal and lateral velocities
  float y_pos; // Longitudinal position within the lane
  int lane; // the lane number
  int task; // Current task
  int task_state; // The current state of the current task
};
```

Note that in contrast to the data members of the command messages, the data members contained in the status message class are those necessary to record all vehicle states of operation, such as position, velocity, current lane, intervehicle spacing, and so on. Note that although not included in the Example 5.5, the current controller type and its parameters can also be a part of the status message. Then we can have a command message, called, for example, DRIVE_SET_CONTROL_PARAM, using which we can switch control algorithms for a particular task or change the controller parameters based on the condition of the vehicle or the road.

As mentioned before, the declaration of command and status messages of any RCS application is similar to those presented here. The only difference is that the members of the messages should be set by the programmer based on the needs of the application.

As with the other NML messages, the RCS command and status messages are encoded and decoded. The NML format function, which consists basically of a switch statement, is used to choose which CMS update function is to be called. Example 5.6 shows the NML format function of the messages of the drive module. Note that there is one entry for each command or status message. If for some reason you add a new message to the message vocabulary of this module, the only change to the format function would be an addition of one more entry to the switch statement.

Example 5.6: Format function for the messages of the drive module of an intelligent vehicle

```
// NML/CMS Format function : driveFormat
int driveFormat(NMLTYPE type, void *buffer, CMS *cms)
{
  switch(type)
    {
    case DRIVE_ACCELERATE_TYPE:
      ((DRIVE_ACCELERATE *) buffer)->update(cms);
      break;
    case DRIVE_CHANGE_LANE_TYPE:
      ((DRIVE_CHANGE_LANE *) buffer)->update(cms);
      break;
    case DRIVE_DECELERATE_TYPE:
      ((DRIVE_DECELERATE *) buffer)->update(cms);
      break;
    case DRIVE_FOLLOW_TYPE:
      ((DRIVE_FOLLOW *) buffer)->update(cms);
      break;
    case DRIVE_HALT_TYPE:
      ((DRIVE_HALT *) buffer)->update(cms);
      break;
    case DRIVE_INIT_TYPE:
      ((DRIVE_INIT *) buffer)->update(cms);
      break;
    case DRIVE_STATUS_TYPE:
      ((DRIVE_STATUS *) buffer)->update(cms);
      break;
    case DRIVE_TRACK_TYPE:
      ((DRIVE_TRACK *) buffer)->update(cms);
      break;

    default:
      rcs_print("driveFormat: Invalid message. \n");
      return(0);
    }
  return 1;
}
```

To make the discussion complete, we will also present the constructor and the update function for one of the messages of the drive module of the intelligent vehicle. The constructor of the message basically calls the base class' constructor (i.e., the constructor of `RCS_CMD_MSG` or `RCS_STAT_MSG` based on the type of the message), in order to allocate memory and to initialize the base class members of the message. Then it initializes the application-dependent members that we added to the message (`optspeed` and `optspacing` in Example 5.7). The update function is a function which calls the CMS methods for encoding and decoding the members of the message in a neutral format. Example 5.7 shows the implementation of the constructor and the update function of `DRIVE_FOLLOW` command message. Implementation of the others is similar. The only important issue is that in the constructor, the appropriate base class constructor should be called, the new data members should be initialized to start from a known state, and in the update function, the CMS update should be called for every member of the message in the order of declaration of the members.

Example 5.7: Constructor and update function for the command message `DRIVE_FOLLOW`

```
//  Constructor for DRIVE_FOLLOW
DRIVE_FOLLOW::DRIVE_FOLLOW()
  : RCS_CMD_MSG(DRIVE_FOLLOW_TYPE,sizeof(DRIVE_FOLLOW))
{
  optspeed = (float) 0;
  optspacing = (float) 0;
}


//   NML/CMS Update function for DRIVE_FOLLOW
void DRIVE_FOLLOW::update(CMS *cms)
{
  cms->update(optspeed);
  cms->update(optspacing);
}
```

This completes the discussion of the messages. In the next section we discuss the NML communication channels, which are used for communicating messages between the modules in an RCS application.

5.3 Setting NML Communication Channels

In this section we first describe how you can set an NML communication cannel using the `NML` base class of the RCS library and later we introduce the two classes `RCS_CMD_CHANNEL` and `RCS_STAT_CHANNEL` which are derived from the `NML` base class and serve as generic NML channels for transferring command and status messages.

5.3.1 Creating an `NML` Object

You can set an NML communication channel basically by creating an `NML` object. An NML object is created via the use of the constructors of the class. NML has several different constructors, but most will follow the form

`NML(NML_FORMAT_PTR` *f_ptr*`, char *`*buf*`, char *`*proc*`, char *`*file*`);`

where *f_ptr* is the address of the format function to use, *buf* is the name of the buffer to connect to (this is specified in a configuration file discussed later), *proc* is the name under which to access the buffer (a process name, also discussed later), and *file* is the name of the configuration file. Constructors can be called in the declaration of a variable, or after a `new` statement if the memory for the object is dynamically allocated. The code in Example 5.8 illustrates different ways of creating a new NML object.

Example 5.8: Creating an NML object

```
/* nml_ex2.cc */
#include "rcs.hh"
#include "nml_ex1.hh"

main()
{
    /* NML( format function, buffer name, process name, configuration
       file ) */
    NML example_nml(ex_format, "ex_buf1", "ex2_proc", "ex_cfg.nml");
    NML *example_nml_ptr;
    example_nml_ptr = new NML(ex_format, "ex_buf2", "ex2_proc",
                              "ex_cfg.nml");
}
```

In this example both the object `example_nml` and the object pointed by the NML pointer `example_nml_ptr` are set up to use the format function `ex_format` from the previous examples. This function will be called on arrival of a new message to the channel. The file `ex_cfg.nml` is an NML configuration file in which the buffers `ex_buf1` and `ex_buf2`, and the processes `ex_proc1` and `ex_proc2` are defined. We discuss configuration files in more detail in Chapter 7.

As in the case of command and status messages, it is a good idea to create generic NML command and status channel base classes which will deal with the corresponding communication between the modules and will pass command and status messages, respectively. The RCS library contains such classes, named `RCS_CMD_CHANNEL` and `RCS_STAT_CHANNEL`. We discuss them next.

5.3.2 Command and Status Channels

The `RCS_CMD_CHANNEL` and `RCS_STAT_CHANNEL` classes are generic classes which are for transmitting command and status messages, respectively. They are derived from the `NML` base class and are a more intuitive and convenient way of setting NML channels for communications between the modules in an application. As mentioned before, it is up to the user whether to use these predefined classes or to create user-defined classes according to the requirements of the application. The member functions of the `RCS_CMD_CHANNEL` and `RCS_STAT_CHANNEL` are as shown in Table 5.2.

Since these messages are derived from the `NML` class, they inherit all the members of this class, plus they have some additional members. The format of the constructors for these messages is

`RCS_CMD_CHANNEL(NML_FORMAT_PTR` *f_ptr*`, char *`*buf*`, char *`*proc*`,`
`    char *`*file*`, int set_to_server = 0);`

Table 5.2: Members of the command and status channels.

Class Name	Members	Purpose
RCS_CMD_CHANNEL	All members from NML plus	
	RCS_CMD_CHANNEL()	Constructor
	~RCS_CMD_CHANNEL()	Destructor
	get_address()	Get address of new message
RCS_STAT_CHANNEL	All members from NML plus	
	RCS_STAT_CHANNEL()	Constructor
	~RCS_STAT_CHANNEL()	Destructor
	get_address()	Get address of new message

```
RCS_STAT_CHANNEL(NML_FORMAT_PTR f_ptr, char *buf, char *proc,
    char *file, int set_to_server = 0);
```

where, as you can see, the first four parameters passed to the constructor are the same as those passed to the constructor of the NML class discussed before. In fact, during creation of an object of the derived class, these parameters are passed to the constructor of the base class NML. The last parameter set_to_server, which is by default 0, is to indicate whether this process (i.e., the process creating the NML channel) is a server for the buffer it connects to. The value 1 indicates it is a server and the value 0 is for indicating nonserver processes. The server processes are responsible for creation of the buffer and the others try to connect to buffers that are already created. This is discussed in more detail later in the book.

The class NML contains a function get_address() which will be discussed later. We mention it here because the derived classes RCS_CMD_CHANNEL and RCS_STAT_CHANNEL redefine it in order to interpret the messages as the generic command and status message classes RCS_CMD_MSG and RCS_STAT_MSG, discussed before. The redefinition and implementation of the get_address() function for the command channel is

```
RCS_CMD_MSG *get_address()
    { return((RCS_CMD_MSG *) NML::get_address()); };
```

and that for the status channel is

```
RCS_STAT_MSG *get_address()
    { return((RCS_STAT_MSG *) NML::get_address()); };
```

which, as you can see, call the get_address() of the base class but interpret the data as command or status message, respectively.

As in the case of RCS_CMD_MSG and RCS_STAT_MSG, the programmer is not required to use RCS_CMD_CHANNEL and RCS_STAT_CHANNEL, and can create all the communication channels using NML. However, these classes will be used in the NML_MODULE class, which is a generic control class, for setting up communication channels between a module and its supervisor and subordinates. For this reason, the programmer may prefer to use the derived classes. Moreover, these classes are more intuitive for module communications in RCS applications.

5.3.3 Sample Command and Status Channels: AHS Example

In this section we return to the AHS example introduced earlier. Consider the supervisor module in the vehicle hierarchy (refer to Figure 5.1). Assume that the diagnostics program for the driver is running on a different computer or backplane. In that case it needs to access the buffers of the supervisor through an NML server. Therefore, we have to establish command and status channels from the NML server to the buffers of the supervisor. Example 5.9 shows how this can be done.

Example 5.9: Command and status channels for the supervisor of an intelligent vehicle

```
// This code is an excerpt from ivhssvr.cc

#include "rcs.hh"        // Common RCS definitions
#include "supervisorn.hh"  // supervisorFormat

// NML Channel Pointers
static RCS_CMD_CHANNEL *supervisor_cmd = NULL;
static RCS_STAT_CHANNEL *supervisor_stat = NULL;

int main (void)
{

  /*  Define variables here
      ...
      */

  // supervisor
  supervisor_cmd = new RCS_CMD_CHANNEL(supervisorFormat, "supervisor_cmd",
                                       "ivhssvr", "ivhs.nml");

  supervisor_stat = new RCS_STAT_CHANNEL(supervisorFormat, "supervisor_sts",
                                         "ivhssvr", "ivhs.nml");

  /*  Do processing
      ...
      */
}
```

Note that the declaration of these channels is very similar to that of NML. You provide as parameters to the constructor a format function to distinguish the messages going through the channel, the name of the buffer to connect to, the name under which to connect to the buffer, and the name of the NML configuration file where the buffers and processes are defined.

Below we discuss features of NML such as reading data from and writing data to a buffer. Since `RCS_CMD_CHANNEL` and `RCS_STAT_CHANNEL` inherit these features of the `NML` class, there will be no difference in reading from a channel defined as the base class or as a derived class. Therefore, we provide a general discussion only on NML.

5.4 Reading and Writing NML Data

Recall that RCS provides functions that allow the user to write to and read from local and remote buffers. By examining many of the communications interfaces, one can find two functions that are similar to the UNIX read and write commands. General read/write commands contain three main parameters: an identifier of the source of the read (or destination of the write), a pointer to a buffer, and the size of the message to be communicated. The read functions cause problems in that the receiver must create a buffer large enough to store the message without having any way of knowing the size of the message to begin with. NML gets around this problem by providing access to a memory buffer that was user configured to a specific size (via an NML configuration file, discussed later in Chapter 7) that will contain a copy of the incoming message after a read operation. Here, first we discuss the basic read and write functions, then explain the general read/write procedure, and after that introduce some additional read and write functions of NML.

5.4.1 Basic Read and Write Functions

C++ allows us to simplify the interface of the write and read functions considerably. The member function to perform a read is

```
NMLTYPE NML::read()
```

which returns 0 if the buffer has not been written to since the last read, -1 if an error occurred, or the type id of the message received if the buffer contains new data (Example 5.10. If the read is successful, the message currently in the global CMS buffer will be copied into a local buffer for this process (the local buffer is specified in a configuration file). The address of this local buffer is available through the member function `get_address()`. If the buffer is encoded, the format function will be called to return the message to native format. The message should be some user-defined type derived from `NMLmsg`. The member `NML::error_type` can be examined to see the cause of `NML::read` returning -1. If queuing is enabled on this buffer, this read will remove the message from the queue so that other processes that are reading from this buffer will see the next message on the queue and potentially miss this one. To obtain the address of the local buffer that the data is stored in, you can use the function

```
NMLmsg *NML::get_address()
```

which returns a pointer to the NML data stored during an `NML::read()` operation.

The write functions are overloaded to accept both a pointer to a message or a reference to a message. The functions are

`int NML::write(NMLmsg &`*nml_msg*`)`	Reference version
`int NML::write(NMLmsg *`*nml_msg*`)`	Pointer version

Both return 0 if successful or -1 otherwise. In both functions *nml_msg* should be a pointer or reference to an object of some defined type derived from `NMLmsg` (e.g., the `EXAMPLE_MSG` class developed above). If the buffer is configured to be in a neutral format, the message will be encoded before it is written to the CMS buffer. It is important to note that the write functions overwrite the message currently in the buffer if queuing is not enabled. Once again, the member `NML::error_type` can be examined to see the cause of `NML::write` returning -1.

Example 5.10: Reading from and writing to an NML channel This example presents an example code which reads and writes our example message `EXAMPLE_MSG` from memory.

```
/* nml_ex3.cc */
#include "rcs.hh"
#include "nml_ex1.hh"

main()
{
    RCS_TIMER timer(0.1);
    NML example_nml(ex_format, "ex_buf1", "ex3_proc", "ex_cfg.nml");
    EXAMPLE_MSG *example_msg_ptr;
    int quit=0;

    /* Write the message to buffer ex_buf1 as configured in ex_cfg.nml */
    example_msg->f = 123.456;
    example_msg->c = 'c';
    example_msg->i = 99;
    example_nml->write(example_msg);

    while(!quit)
        {
            switch(example_nml.read())
                {
                case -1:
                    rcs_print( "A communications error occurred.\n");
                    quit = 1;
                    break;

                case 0:
                    /* The buffer contains the same message */
                    /* you read last time. */
                    break;

                case EXAMPLE_MSG_TYPE:
```

```
                    example_msg_ptr =
                      (EXAMPLE_MSG *)example_nml.get_address();
                    rcs_print(" We have a new example message. \n");
                    rcs_print(" The value of its members are:\n ");
                    rcs_print(" f=%f, c=%c, i=%d\n ",
                              example_msg_ptr->f,
                              example_msg_ptr->c,
                              example_msg_ptr->i);
                    quit = 1;
                    break;
                }
            timer.wait();
        }
}
```

In this sample routine we create an NML object called `example_nml` that uses the format function `ex_format` created in a previous example. The other parameters of the NML constructor are the name of the buffer this process will access (`ex_buf1`), the process name (`ex3_proc`), and the name of the configuration file (`ex_cfg.nml`) that sets up the size of the buffer and which processes will access the buffer as was discussed in the preceding section. Also, the program uses the `EXAMPLE_MSG` message type that was created in Section 5.2. In attempting to read the NML buffer, we must check for each case that the `NML::read()` function returns by setting up a `case` structure. This example also uses the `RCS_TIMER` class and the `rcs_print` function described in the Chapter 8.

The read/write procedure for `RCS_CMD_CHANNEL` and `RCS_STAT_CHANNEL` is similar; therefore, we do not present an example of them here. In Chapter 6 we present examples of reading and writing messages to the command and status channels by the control module `NML_MODULE`.

5.4.2 General NML Read/Write Procedure

Figure 5.2 shows the typical overall read/write access operation of an NML process that accesses two buffers. The NML read operation calls the appropriate CMS communications functions, which call the `update()` functions to decode the message from the buffer. Once the unique identification number of the message currently in the buffer is determined, a user-written format function is called that simply tells NML what update function to call based on the type of message that is being read from the buffer. Once the decoding is complete, the CMS read causes the resulting message to be written to local memory of the process where it can be accessed. The application-specific code then can continue executing. At the end of the cycle, the application code generally needs to write its own messages back to the buffer to communicate with remote processes. During the write operation, the NML `write()` function is called, which first calls the format function which selects the appropriate `update()` function for encoding (recall that each message has its own `update()` function—the correct one must be selected when writing/reading messages, and the format function accomplishes this). The message-specific update function calls the lower-level CMS update functions for encoding the individual members of the message. Once encoding is complete, CMS communication mechanisms are called that actually access the buffers and perform the write.

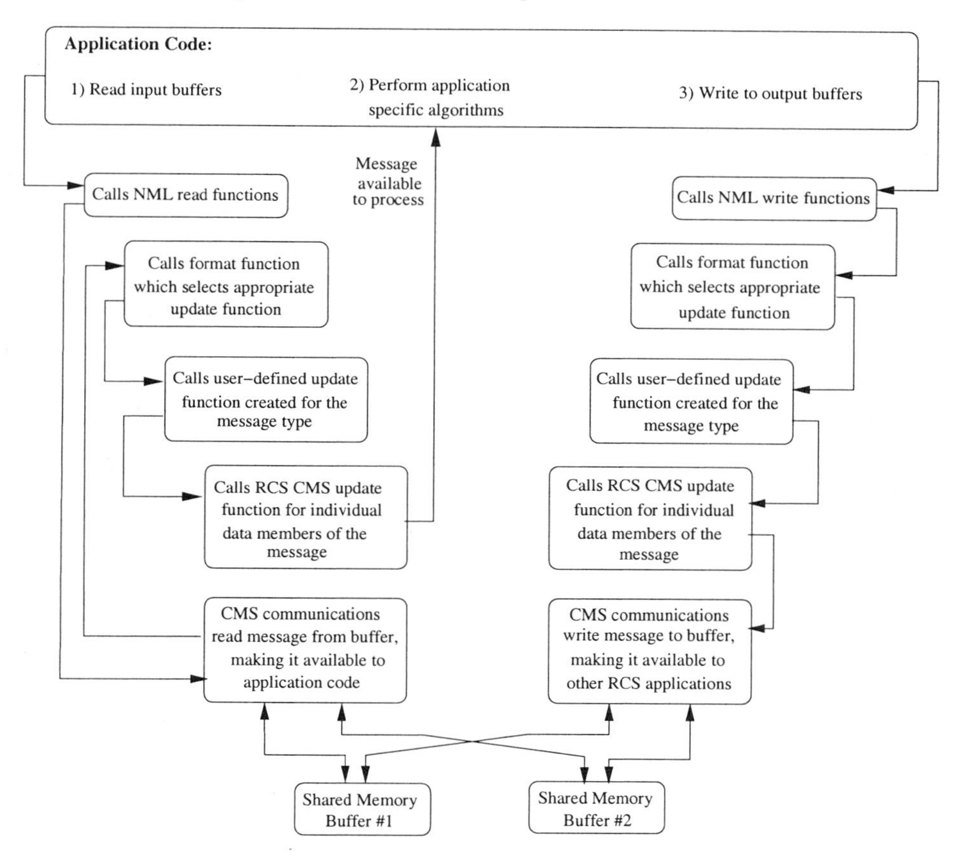

Figure 5.2: Typical operation of a process using NML.

5.4.3 Additional Read/Write Functions

We can determine if a buffer has been written to without actually affecting any of the flags by using the function

```
NMLTYPE NML::peek().
```

This function returns 0 if the buffer has not been written to since the last read, -1 if an error occurred, or the type id of the message received if the buffer contains new data. It works exactly the same way as read except that the flag that lets others know when the buffer is read is not changed, and if queuing is enabled, the message is not removed from the queue. This could be useful if you need to monitor a buffer without letting other processes using the buffer know.

The member NML::error_type can be examined to see the cause of NML::peek returning -1.

An additional write command which checks whether that message has been read is also available to the programmer and is defined as follows:

```
int NML::write_if_read(NMLmsg &nml_msg)        Reference version

int NML::write_if_read(NMLmsg *nml_msg)        Pointer version
```

These functions combine the operations of writing and checking if a buffer has been read. It checks to see if the buffer has been read, and if it has, then it writes the message into the buffer just as write would. If it has not been read, then it returns -1. This function returns a 0 if successful and a -1 otherwise. Since there is only one access to the buffer, there is no way for another process to write into the buffer between the check and the write. The member NML::error_type can be examined to see the cause of NML write_if_read returning -1.

Checking If Data Has Been Read

If queuing is not enabled, then a flag is kept in every CMS buffer called was_read. Every time a write is performed on the buffer the flag is set to 0. Every time a read is performed on the buffer the flag is set to 1. The check_if_read function just returns the value of that flag. To avoid overwriting a buffer that has not been read yet, it is better to use the write_if_read function. NML::error_type contains information as to why NML::check_if_read returned -1.

If queuing is enabled, then the check_if_read function returns 1 only if all of the messages in the buffer have been read (meaning that the queue is empty). The prototype of this function is

```
int NML::check_if_read()
```

It returns 0 if the buffer contains a message that has never been read, 1 if the buffer contains a message that has been read at least once, or -1 if an error occurred that prevented NML from determining whether the buffer has been read.

Clearing a Buffer

You may want to clear a buffer to preempt previously sent messages still in the queue or to ensure that residual data in a buffer is not mistaken for NML messages. The function that accomplishes this is

```
int NML::clear()
```

which returns 0 if the buffer was successfully cleared and -1 if an error occurred.

Converting an NML Message to a String

It is occasionally helpful to be able to display the contents of any NMLmsg in a string. To accomplish this you will need an NML object which was initialized with a format function that handles your message type. Then passing the pointer or reference to this object as an argument to the function

```
const char * NML::msg2str(NMLmsg *)

const char * NML::msg2str(NMLmsg &)
```

will convert the contents of the message to a displayable string. This function returns a pointer to a string with each member of the NML message converted to a string and separated with commas if successful or `NULL` otherwise. The first two members will be the type and size of the message. The string may be cleared on the next call to `read`, `write`, `check_if_read`, `peek`, `write_if_read`, `clear`, or `msg2str` with that NML object, or when that NML object is deleted, so the string should be displayed or copied before any of these operations occur.

5.5 Error Handling

When the NML member functions cannot perform their tasks, they try to provide developers with some information that may allow them to resolve the problem. This information is available in several forms, such as by setting a return value to an error code or by setting the `NML::error_type` data member to an appropriate value. In addition to these, NML has a member function, called `valid()`, for checking the validity of the object initialized.

5.5.1 Error Types of NML Read/Write Functions

The NML functions `NML::read()`, `NML::write()`, `NML::write_if_read()`, and `NML::peek()` return -1 if an error occurred. On the other hand, the function `NML::get_address()` returns `NULL` if an error occurs.

Messages are printed to character display devices or stored in a linked list with the `rcs_print_error` facility. Often, several messages are issued for the same error, because if an error occurs at a low level, the low-level function will print an error and return a value indicating an error to a higher-level function, which may then also print an error. This allows the user to see the details available at the lower level and the context available at the higher level.

`NML::error_type` is a variable set by NML functions that fail. It may have only the following values:

- `NML_NO_ERROR` – No error was recognized.
- `NML_INVALID_CONFIGURATION` – A problem with the configuration file is indicated.
- `NML_BUFFER_NOT_READ` – Operations like `write_if_read` will succeed only if the message currently in the buffer has been read.
- `NML_TIMED_OUT` – An operation timed out.

- NML_FORMAT_ERROR – Could indicate that there was a problem with the user-defined format and update functions, that the size of the buffer is not large enough for one of the messages, or that a message was received that no format function recognizes.
- NML_NO_MASTER_ERROR – Something needs to be initialized by the process that is configured to be the buffer master. Check that a master is configured and running.
- NML_INTERNAL_CMS_ERROR – A CMS operation failed for a reason not previously given.

Example 5.11 shows a coded C++ segment that shows how you can check the error status of an RCS application.

Example 5.11: Checking error status This example shows how to check the error status of an RCS application.

```
/* nml_ex6.cc */

#include "rcs.hh"
#include "nml_ex1.hh"

/* This example prompts the user when NML times out to see if it
   should try again. */

main()
{
    NML example_nml(ex_format, "ex_buf1", "ex6_proc", "ex_cfg.nml");
    EXAMPLE_MSG *example_msg_ptr;
    char input_array[10];

TRY_AGAIN:

    switch(example_nml.read())
        {
        case -1:
            if(example_nml.error_type == NML_TIMED_OUT)
                {
                    rcs_print("NML timed out\n");
                    rcs_print("Do you want to try again? (y/n)");
                    gets(input_array);
                    if(input_array[0] == 'y')
                        goto TRY_AGAIN;

                }
            break;

        case 0:
            /* The buffer contains the same message you read last time. */
            break;

        case EXAMPLE_MSG_TYPE:
            example_msg_ptr = (EXAMPLE_MSG *)example_nml.get_address();
            /* We have a new example message. */
```

```
                break;

        }
}
```

5.5.2 Checking If an NML Object Is Valid

After creating an NML object you can check to make sure that it is valid. The function that determines validity of an NML object is

```
int NML::valid().
```

`NML::valid()` returns 0 if the object was not properly constructed or if an error has occurred which is severe enough so that it is unlikely that any of the other NML operations on this object will succeed, or 1 if everything seems to be in order.

Example 5.12 shows how you can check the validity of the NML auxiliary, command, and status channels created for the intelligent vehicle supervisor. This code is an excerpt from the NML server for the buffers in the controller. The function `InitNML()` is a function that initializes all the NML channels and is called within the main function of the server. The initialization of the channels to the other buffers is not shown since it is similar to the ones shown.

Example 5.12: Validity check for command and status channels for the supervisor of an intelligent vehicle

```
// This code is an excerpt from ivhssvr.cc

// Include Files
#include "rcs.hh"         // Common RCS definitions
#include "supervisorn.hh"  // supervisorFormat
#include "attn_to_supn.hh"  // attn_to_supFormat

// NML Channel Pointers
static RCS_CMD_CHANNEL *veh_superv_cmd = NULL;
static RCS_STAT_CHANNEL *veh_superv_stat = NULL;
static NML *attn_to_sup= NULL;

static int InitNML()
{
  // supervisor
  supervisor_cmd = new RCS_CMD_CHANNEL(supervisorFormat, "supervisor_cmd",
                                       "ivhssvr", "ivhs.nml");
  if(NULL == supervisor_cmd)
    return -1;
  if(!supervisor_cmd->valid())
    return -1;

  supervisor_stat = new RCS_STAT_CHANNEL(supervisorFormat, "supervisor_sts",
                                         "ivhssvr", "ivhs.nml");
  if(NULL == supervisor_stat)
```

```
      return -1;
    if(!supervisor_stat->valid())
      return -1;

    attn_to_sup = new NML(attn_to_supFormat, "attn_to_sup",
                          "ivhssvr", "ivhs.nml");
    if(NULL == attn_to_sup)
      return -1;
    if(!attn_to_sup->valid())
      return -1;
}
```

In this example, when an NML object is initialized, if the process is declared in the configuration file as a master for the buffer it tries to create the buffer. On the other hand, if it is not the master, then it tries to connect to the specified buffer. If the process is configured as a master and is unable to allocate memory to the buffer, an error occurs and the object is nonvalid. Similarly, if the process is not a master and tries to connect to a buffer, which was not created by another process before, an error occurs. This forces the function to return with an error code of -1.

5.6 Spawning and Killing NML Servers

As we mentioned before, NML servers allow remote processes to access local buffers. They read and write, and encode and decode, the messages in the buffer locally on behalf of the remote processes. The code for the servers has already been included in the RCS library, but you must still start and stop them. There are several ways for you to control when servers are spawned and killed. One way is to dedicate to the servers an executable that will not perform any other operation and run the servers using a function which will not return, and the other way is to start and stop them within the program as required by the needs and allow for the calling program to do some other processing, too.

5.6.1 Server Program Without Application Code

If we do not need the executable for the NML servers to perform any other operation, we can create NML objects, initialize NML servers, and let them run forever. In order to initiate the NML servers, you can use the command

```
void run_nml_servers()
```

Each time an NML object is created it is added to a global list. In operating systems like LynxOs and SunOs that use heavyweight threads, each process has its own list. In operating systems like VxWorks that use lightweight threads, the list is shared by all processes currently running that use NML. The function `run_nml_servers` reads the lists, checks for the server configuration flag, and groups the buffers with the same *remote procedure call* (RPC) number. For each different RPC program number a server is spawned to handle requests for the group of buffers with that RPC number. If all of the RPC numbers were the same, the current process would become the server for all the buffers on the list that have a nonzero server configuration flag. This function will not return.

Example 5.13 shows the use of the `run_nml_servers` function. It is an excerpt from the code for the NML server for the buffers of the intelligent vehicle. Note that it calls the `InitNML()` function that initializes and checks the validity of all the NML channels (refer to Example 5.12, where we showed part of that function).

Example 5.13: NML servers for the intelligent vehicle

```
// This code is an excerpt from ivhssvr.cc

// Include Files
#include <stdlib.h> // exit()
#include "rcs.hh"       // Common RCS definitions

int main(void)
{
  // Print error to the screen
  set_rcs_print_destination(RCS_PRINT_TO_STDOUT);

  if(InitNML() < 0)
    {
      DeleteNML();
      rcs_exit(-1);
    }

  rcs_print("\nRunning servers . . .\n\n");
  run_nml_servers();
}
```

The function `rcs_print_destination` is a function that specifies the destination for the outputs of the print function. The `rcs_print` and `rcs_exit` functions are simply the `print` and `exit` functions of the RCS library. All these functions are discussed in Chapter 8.

5.6.2 Server Program with Application Code

Sometimes we may need to do some more processing in the function which starts the NML servers. Since `run_nml_servers()` does not return, we need other means to start and stop NML servers. Two functions available for starting and stopping NML servers are

```
void nml_start()
```

and

```
void nml_cleanup().
```

The `nml_start` function works like `run_nml_servers` except that it may spawn an additional process(es) so that it will return. The `nml_cleanup` function deletes all NML_SERVER objects and all NML objects so that all servers that were spawned will be stopped with SIGINT.

If we had needed the server program in Example 5.13 to perform some other processing while the servers are running, we could have implemented it as in Example 5.14.

Example 5.14: NML servers for the intelligent vehicle with an application code

```
// Include Files
#include <stdlib.h> // exit()
#include "rcs.hh"       // Common RCS definitions

int main(void)
{
  // Print error to the screen
  set_rcs_print_destination(RCS_PRINT_TO_STDOUT);

  if(InitNML() < 0)
    {
      DeleteNML();
      rcs_exit(-1);
    }

  rcs_print("\nRunning servers . . .\n\n");
  /* Spawn the servers and continue */
  nml_start();

  /* ...... */
  /* Do some processing */
  /* ...... */

  /* Kill the servers and close the NML channels */
  nml_cleanup();
}
```

5.7 User Command Utilities

There are several NML utilities that can be used directly by users without any additional programming. These include testing for the existence of NML buffers, determining the performance of NML on a particular system, and removing unwanted buffers. We describe these utilities next.

5.7.1 Testing for the Existence of NML Buffers

You can test for the existence of an NML buffer using the function

```
nmltest [config_file local_host].
```

The nmltest program reads the config_file and attempts to connect to every buffer in the file. If it succeeds, the buffer exists. If it does not, either the master for the buffer was not started, or the server was not started, or the configuration file contains an error. It attempts to use the local protocol for buffers on the local_host. If you do not specify the config_file or local host, it will prompt you for them. It also reports the type and size of message, if any, in the buffer and some other information which may be useful.

5.7.2 Determining the Performance of NML on a Particular System

To determine the performance of NML, use the programs

```
nmlperf [config_file local_host master iterations increments
    detailed_output display_mode]

perfsvr [config_file local_host]
```

The performance of NML varies depending on the system you are running under and the type of protocol used by CMS. The nmlperf program connects to every buffer in config_file. It uses the local protocol if the buffer is on local_host. It initializes the buffer if master equals 1. It writes and reads the buffer with messages from a minimum size to the size of the buffer for iterations times with increments of different sizes. If detailed_output equals 1, then it will display the time required for every read and write. The parameter display_mode can be "B" for bytes-per-second or "M" for messages-per-second.

If you do not specify all the parameters, then you will be prompted for them. The perfsvr program is run on a remote machine to allow testing of remote protocols.

5.7.3 Removing Unwanted Buffers

Sometimes it is desirable to remove buffers that are not being used in order to "clean up" the system. The function

```
nmlclean [config_file local_host]
```

performs this task.

Occasionally, severe errors cause programs to exit without deleting the NML buffers that they create. The nmlclean program attempts to free all of the operating system resources associated with buffers in the config_file on the local computer (local_host) and to kill the NML servers that are running.

In this chapter we presented the basic functions of NML together with some illustrative examples. In the next chapter we discuss the RCS control module, which is the main computing module and building block in the RCS hierarchy.

Chapter 6

RCS Control Module

In this chapter we discuss the RCS control module, which is a generic control module. Its development is inspired from the basic building block of the RMA discussed in Chapter 3 (refer to Figure 3.1). on page 67). We know that RCS has a hierarchical structure. Every control module in the hierarchy accepts commands from its supervisors and sends commands to its subordinates, while sensory processing data is passed from subordinates to their superior modules. RCS has cyclic processing. In other words, the operations are done on cyclic intervals by the control module. We know that every module can have its own timing depending on the requirements of the application. In general, the processes in the lower levels of the RCS hierarchy become faster than those in the higher levels (refer to the timing diagram of the RMA discussed in Chapter 3).

The operation of a RCS control module is similar to that of a conventional discrete time control system. There in each sampling interval, the controller reads the measurable signals, calculates the control value based on some algorithm, and outputs it to the plant. In the RCS control module, we have similar processing. In other words, each cycle time the module performs some predefined operations in a predefined order. In particular, in each cycle a module reads its sensory data, checks the status of its subordinates, checks for commands from its supervisors, then based on all these, it updates its database, generates a future plan, chooses the next operation to be performed, sends appropriate commands to its subordinates (if necessary), outputs required signals (if any), reports its status to its superior, and waits till the end of the cycle.

Consider, for example, the maneuver module in the AHS problem discussed before. Every cycle time it reads the information from the attention and drive modules such as incoming messages, `optspeed`, `optsize`, speed of the vehicle, conditions of the road, conditions of the engine, conditions of the brakes, and so on, and updates its database. Next, based on the data read and the maneuver requests from the supervisor module, it sends commands to the drive module such as `ACCELERATE`, `CHANGE_LANE`, `DECELERATE`, and so on, or to the attention module such as `SEND_REQ`, `BROADCAST`, and so on.

These observations suggest the idea of development of a generic RCS module base which will implement the application-independent skeleton of the process described. Since every module needs to establish some means of communication with other modules, adding the tools for communication with its supervisors and its children (and maybe other modules) will enhance the module base. This leads to development of a base class for RCS control modules called `NML_MODULE`. The `NML_MODULE` base class, referred to also as an *RCS template* (*note:* this is *not* referring to a C++ `template`), primarily uses NML for communications. It provides a consistent structure for each control module, and performs operations common to all modules, such as initializing NML, measuring performance, and checking for status of lower-level modules and for new commands from upper modules. You can use the `NML_MODULE` base class by deriving new application-specific control module classes from it.

In general, each NML module is associated with two buffers—a command buffer (readable to the module) and a status buffer (writable to the module)—and has both parent modules and subordinate modules. Parents have access to the subordinates' command and status buffers. Generally, the parents obtain status information from the subordinate's status buffer and maintain the ability to command the subordinates with the ability to write command messages to the command buffer. This is illustrated in Figure 6.1. Module 1, at the top of the hierarchy, has the ability to command modules 2 and 3 by sending command messages to modules 2 and 3's command buffers. Likewise, module 1 can gather status information by reading modules 2 and 3's status buffers. Each node of the hierarchy behaves in this fashion, mimicking the hierarchical structure of the RMA. The operator interface can send commands to the upper-level module, where initial task decomposition occurs.

6.1 NML_MODULE Members and Operation

In this section we describe the basic functions and data members of the RCS generic module class, `NML_MODULE`. Some of the member functions are declared as virtual, allowing for overloading so that the derived class can be designed for a specific application.

The operation of a single cycle of an NML module is shown in Figure 6.2. Items boxed with a dashed line are operations which are performed by functions already available in the RCS library; they do not need to be written by the user. Those items in solid lines are application-specific code that the user needs to develop for the system (these are usually the virtual functions discussed above). The main loop is implemented once every sampling period of this particular module. To "activate" the controller module, a simple call to the NML modules `controller()` function is necessary. This function, in turn, calls functions that start the timing for the cycle (for diagnostics purposes) and read the input buffers of the particular module (e.g., the command buffer and the subordinate status buffers). Then application-specific code is implemented in the order given above, for the preprocessing, decision-processing, and postprocessing functions. The `controller()` cycle ends with the writing of any necessary messages to the

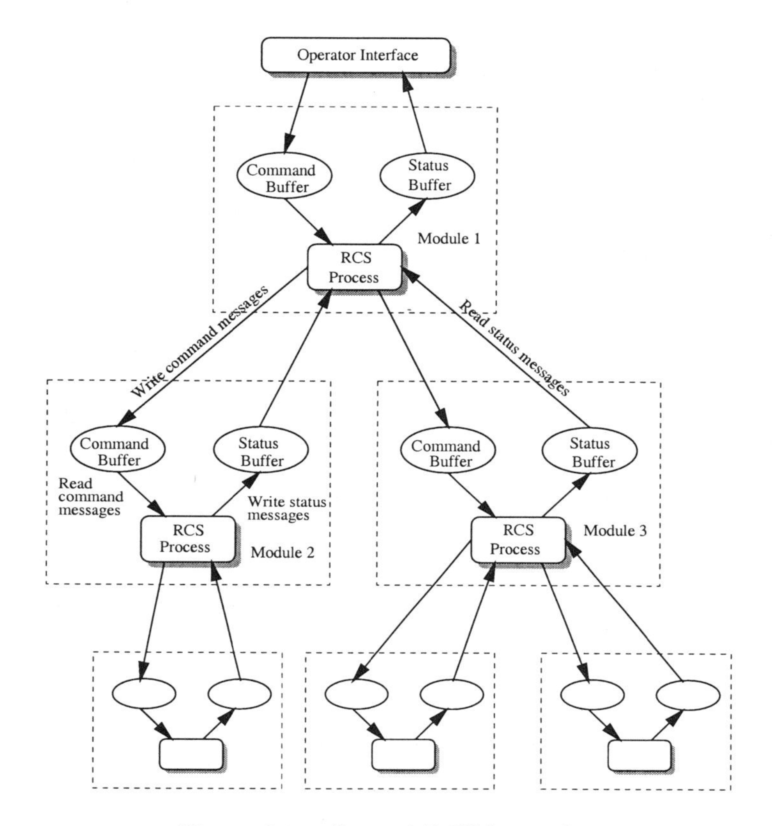

Figure 6.1: General RCS hierarchy.

output buffers, where, for example, commands may be passed to subordinate modules.

Note that since all modules of a hierarchy can act as independent processes, the sampling times can be different for each module. Furthermore, each module can act on separate computers (along separate backplanes, running different operating systems, or even located in remote locations to each other), communicating via a local area (or wide area) network.

Most of the functions of direct consequence to the RCS designer are shown in Figure 6.2. In actuality, there are numerous other members in the `NML_MODULE` base class that more advanced programmers (or more complex systems) may utilize. The following list contains a more detailed description of the functions introduced as well as many additional members (however, the list is not exhaustive):

- `void controller()`
  The `controller` calls the other functions which implement the control algorithm (refer to Figure 6.2). It should be called every cycle from the main loop. It reads the NML input buffers, calls the functions `PRE_PROCESS`,

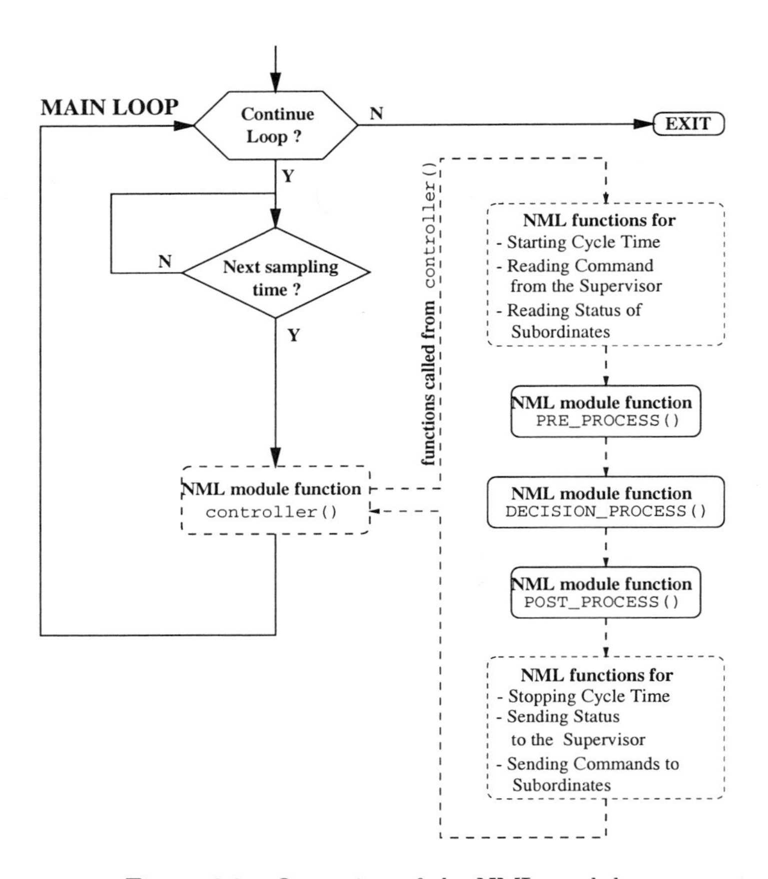

Figure 6.2: Operation of the NML module.

DECISION_PROCESS, and POST_PROCESS, writes to the NML output buffers, and updates the performance metrics. Its code is already implemented in the RCS library; therefore, the programmer is not responsible for programming the body of the function itself.

There are several ways to establish the sampling time for the NML_MODULE loop. One possible (and perhaps the easiest) way is to use the RCS_TIMER provided by the RCS library. The RCS_TIMER class is discussed in more detail in Chapter 8.

- virtual void PRE_PROCESS()
  The PRE_PROCESS function is called by the controller function before calling the DECISION_PROCESS function. Its purpose is to perform simple conversions and sensory processing and read any auxiliary inputs before evaluating the situation and making any decisions. The operations performed in PRE_PROCESS are those which need to be done every cycle regardless of the current command or state. Since, in general, the operations

to be performed within this function are application-dependent, the user usually overloads it and implements the code based on the needs of the application.

- `virtual void POST_PROCESS()`
  The `POST_PROCESS` function is called by the controller function after calling the `DECISION_PROCESS` function. Its purpose is to perform simple conversions and output processing and write any auxiliary outputs that need to be done every cycle regardless of the current command or state. Similarly to the `PRE_PROCESS`, the programmer overloads this function and implements the code based on the application at hand.

- `void DECISION_PROCESS()`
  The `DECISION_PROCESS` function is called by the controller function after `PRE_PROCESS` and before `POST_PROCESS`. Its purpose is to call one of the command functions based on the current command type. Command functions are functions which implement the control algorithm associated with a given command or task request. In other words, there are predefined tasks (operations or commands) that each module can perform and there is a function associated with each task. For example, accelerating and changing lanes are performed by two different functions in the intelligent vehicle discussed before.

  In general, the `DECISION_PROCESS` function is implemented as a simple `switch` statement, where each `case` corresponds to a different command received and calls the corresponding function.

- `int addSubordinate(RCS_CMD_CHANNEL *, RCS_STAT_CHANNEL *)`
  The `addSubordinate` function is normally called from within the module's constructor to set an NML channel to send commands to a particular subordinate (i.e., a module on a lower level which accepts commands from, or passes information to, this module). Moreover, it sets an NML channel also to check for status updates from that subordinate every cycle. Each new subordinate is associated with a unique number for identification which is used in calls to `sendCommand`. Moreover, an array of memory, called `statusInData`, is allocated for reading the status of the subordinates. The id of the subordinate is used as an index for the position of its status in the array. The `addSubordinate` function returns the subordinate's number.

  Note that the function uses the generic NML command and status channels `RCS_CMD_CHANNEL` and `RCS_STAT_CHANNEL` for establishing communications.

- `void sendCommand(RCS_CMD_MSG *, int` *subordinate_number*`);`
  The `sendCommand` function sends the command specified by the first argument to the subordinate specified by the second. A subordinate is a control module situated on the lower level of the control hierarchy. Each time a new command is sent, the serial number of the commands is incremented

and the flag `NEW_COMMAND` is set. This allows the subordinate to see that a new command has arrived and to perform processing accordingly.

- `void modifyCommand(RCS_CMD_MSG *, int` *subordinate_number*`)`
  The `modifyCommand` function sends the command specified by the first argument to the subordinate specified by the second in the same way as `sendCommand`, except that the serial number is not incremented. The subordinate will not see this as a new command, but rather continue processing the old command but with the new parameters, which are specified by the new command message.

  To illustrate, consider the intelligent vehicle. Assume that the drive module is currently implementing the `TRACK` command, while the `optspeed` changed. Then the maneuver module, instead of sending a new `TRACK` command using `sendCommand`, may use `modifyCommand` with the new `optspeed` since the actual task did not change.

- `void setCmdChannel(RCS_CMD_CHANNEL *)`
  The `setCmdChannel` function is normally called from within the module's constructor to set an NML channel to check for commands every cycle. In other words, it sets a channel through which the upper-level modules will be sending commands to the given module.

- `void setStatChannel(RCS_STAT_CHANNEL *, RCS_STAT_MSG *)`
  The `setStatChannel` function is normally called from within the module's constructor to set an NML channel and the address of a message to be written out every cycle with status information for the module's superior.

- `void setErrorLogChannel(NML *)`
  The `setErrorLogChannel` function is normally called from within the module's constructor to set an NML channel where error messages will be written when `logError` is called.

- `int loadDclock(double` *seconds*`)`
  Initializes a count-down delay clock which will expire, causing `checkDclock` to return 1 after the number of seconds specified. Fractions of a second are allowed. The actual resolution depends on the accuracy of the system clock.

- `int checkDclock()`
  Returns 1 if the count-down delay timer has expired, 0 otherwise.

- `int STATE_MATCH(int` *state*`, int` *conds*`)`
  The `STATE_MATCH` function is used within the state tables and allows the current line in the state table to be tracked in the RCS diagnostics tool. It returns a nonzero integer if the current state is equal to the first argument and the second argument evaluates to a nonzero number; otherwise, it returns zero.

- `void stateNext(int` *state*`)`
  This function is used within the state tables to switch the state. On the next cycle, after calling this function with a new state value, a new `if` statement in the state table will be true, allowing the current line to be tracked in the RCS diagnostics tool.

- `void logError(const char *`*fmt*`, ...)`
  The `logError` function logs a message to the NML channel specified with `setErrorLogChannel`. The message will contain a string created from the format and a variable number of arguments using the `printf` conventions. The message can be retrieved and displayed by the RCS diagnostics tool, discussed in detail in Chapter 9, or a custom *graphical user interface* (GUI).

- `void logText(const char *`*fmt*`, ...)`
  The `logText` function logs a message to the NML channel specified with `setErrorLogChannel`. The message will contain a string created from the format and a variable number of arguments using the `printf` conventions. The message can be retrieved and displayed by the RCS diagnostics tool or a custom GUI. This can be used for supplying the operator with information about a particular tool or process, and so on.

- `void requestDisplay(const char *`*url*`)`
  The `requestDisplay` function logs a message to the NML channel specified with `setErrorLogChannel`. The message will contain a string with the *universal resource locator* (URL). The RCS diagnostics tool or a custom GUI can have a Web browser open automatically to the URL, where extended prewritten instructions, figures, or images for the operator would be located.

- `int status`
  This variable is for holding the current status of this module. It is always sent to the supervisor or parent of this module. It should normally be equal to one of the following symbolic constants:

  - `RCS_EXEC`
  - `RCS_DONE`
  - `RCS_ERROR`

  The rest of the data sent to the superior can be accessed at `RCS_STAT_MSG` `*`*statusOutData*, but users of the RCS design tool, which is discussed in detail in Chapter 10, will usually find it more convenient to use the variable that is created automatically, of the form `<module_name>_status`. This variable points to the same location but has a module-specific type.

- `RCS_CMD_MSG *`*commandInData*
  The variable *commandInData* points to an area of memory where commands are placed when they are received. Each cycle time, the module

checks this area in the DECISION_PROCESS function and calls the appropriate function based on the type of the command received. It passes to the called function the message as a parameter.

- RCS_STAT_MSG ***statusInData*
The *statusInData* variable points to an array of buffers where the statuses of subordinates are placed when they are received. The index for the status data from a particular subordinate is the identification number of this subordinate. Users of the RCS design tool will usually find it more convenient to use the variable that is created automatically, of the form <module_name>_status. This variable points to the same location but has a module-specific type.

- RCS_STAT_MSG **statusOutData*
This variable is a pointer to the buffer of data that will be sent to this module's superior. Users of the RCS design tool will usually find it more convenient to use the variable that is created automatically of the form <module_name>_status. This variable points to the same location but has a module-specific type.

These are the basic members of the NML_MODULE class. They are defined considering common functions and data which most modules use. For example, every process performs some sensory processing; then according to the data obtained, it takes an appropriate action (i.e., applies the control algorithm); and finally, saves data or shares it with other modules (e.g., its supervisor) before restart of the cycle. This is the motivation for the functions PRE_PROCESS, DECISION_PROCESS, and POST_PROCESS. Similarly, every module establishes NML communication channels with its superiors and subordinates. It receives commands from its supervisors and reports status to them, and sends commands to, and reads the status of its child modules. Therefore, we have functions such as addSubordinate, sendCommand, modifyCommand, setCmdChannel, and setStatChannel, and data members such as commandInData, statusInData, and statusOutData. Similar arguments lie beyond the definition of the other members of the class.

The RCS application programmer can simply derive the modules from the NML_MODULE and add to it application-specific function and data members. This eases the job and unifies RCS controller development. Note, however, that the programmer needs to overload the functions PRE_PROCESS, DECISION_PROCESS, and POST_PROCESS according to the needs of the application at hand and the command functions defined. These functions are called every cycle by the controller function, which is not normally overloaded. (RCS uses cyclic processing, normally with a constant cycle time for a given module or group of modules set as a parameter to the RCS_TIMER constructor in the main loop. Note, however, that different modules can be set to have different sample times, depending on their needs.)

PRE_PROCESS is normally used for retrieving inputs and performing simple conversions needed every cycle, and often by multiple commands. For example, you might read an encoder input and multiply by a scale factor to convert encoder ticks to millimeters. For more complicated sensor processing, it may be better to develop a separate module just for sensory processing or even an entire hierarchy. It can also be used to read any auxiliary input.

DECISION_PROCESS usually calls one of the command functions based on the current command type. The command functions usually contain state tables made up of multiple "if ... else if ..." blocks. The state tables are designed so that only a small amount of processing needs to occur each cycle and long complex commands are spread over many cycles. The current line in the state table is the first line where the conditions for the if statement are true. This line is highlighted within the RCS diagnostics tool, which allows a human operator to see where the current execution is in the state table.

POST_PROCESS is normally used for writing outputs and performing simple conversions on the outputs needed every cycle by multiple commands. For example, you might multiply a commanded velocity by a scale factor before writing a voltage to a digital-to-analog converter (DAC). It can also be used to write any auxiliary output.

Another issue to mention here is that in the NML_MODULE base class the NMLmsg and NML classes are used only for the errorlog channel and the auxiliary channels. On the other hand, communication with the supervisor and the subordinates of the module is done using the base classes RCS_CMD_MSG and RCS_STAT_MSG for the NML messages, and the base classes RCS_CMD_CHANNEL and RCS_STAT_CHANNEL for the NML channels (see the definition of the function and data members in the class). These classes are derived from NMLmsg and NML, respectively, and were discussed in Chapter 5. Here, we want to mention that they are generic NML message and communication channel classes designed specifically to deal with command and status messages and their communication.

In this section we discussed basic parts of the NML_MODULE class with some examples. In the rest of this chapter we show how you can put all these together and use them for RCS controller development.

6.2 Deriving a New Control Module from NML_MODULE

As was mentioned above, you can derive application-specific control modules from the NML_MODULE class. This is done by declaring a new C++ class inherited from the NML_MODULE class and is done in a C++ header file. Example 6.1 illustrates this derivation for the supervisor module of the intelligent vehicle. (See Figure 4.5 for the modules in the controller for the intelligent vehicle.)

Example 6.1: Header file for the SUPERVISOR_MODULE class derived from NML_MODULE

```
// supervisor.hh -- This C++ header file defines the class SUPERVISOR_MODULE

// Prevent Multiple Inclusion
#ifndef SUPERVISOR_HH
#define SUPERVISOR_HH

// Include Files
#include "rcs.hh"  // Common RCS definitions
#include "nml_mod.hh"  // NML_MODULE definitions

#include "supervisorn.hh" // NML Commands and Status definitions for supervisor
#include "maneuvern.hh"   // NML Commands and Status definitions for maneuver

// auxiliary Input NML Message Files
#include "attn_to_supn.hh" // NML Messages for attn_to_sup

class SUPERVISOR_MODULE: public NML_MODULE
{
public:
  SUPERVISOR_MODULE(int _is_base_class = 0); // Constructor

  // Overloaded Virtual Functions
  virtual void PRE_PROCESS();
  virtual void DECISION_PROCESS();
  virtual void POST_PROCESS();
  virtual void INITIALIZE_NML();

  // Command Functions
  virtual void INIT(SUPERVISOR_INIT *);
  virtual void HALT(SUPERVISOR_HALT *);
  virtual void AUTODRIVE(SUPERVISOR_AUTODRIVE *);

  // Convenience Variables
  SUPERVISOR_STATUS *supervisor_status;
  int maneuver_sub_num;
  MANEUVER_STATUS *maneuver_status;

  // auxiliary Input NML Channels
  NML *ATTN_TO_SUP_CHANNEL; // NML Channel for attn_to_sup
  ATTN_TO_SUP_MSG *attn_to_sup_data; // NML Data for attn_to_sup

private:
  // Add custom variables and functions here.
  int id, pltn_id; // vehicle id and platoon id
  int lane, section; // current lane and highway section
  int optsize, ownsize, pos; // optimum and current platoon size, position in
  float optspeed, speed; // otimum speed, vehicle speed
  int busy; // busy flag
};

#endif  // SUPERVISOR_HH
```

Note that several other C++ header files are included in this file. The file rcs.hh contains definitions for most of the utilities in the RCS library. The base class NML_MODULE is defined in the file nml_mod.hh. The other files contain the NML command and status message class definitions for this module and the subordinates of this module. In particular, the file supervisorn.hh contains definitions of the command message classes SUPERVISOR_INIT, SUPERVISOR_HALT, SUPERVISOR_AUTODRIVE and the status message class SUPERVISOR_STATUS for the supervisor.

The file maneuvern.hh provides the NML messages for the subordinate maneuver module, whereas attn_to_supn.hh provides the definition of the NML messages that will be passed through the auxiliary channel from the attention module to the supervisor module. Recall that an auxiliary channel is a communication channel that is established between processes in the same level of the hierarchy or more than one level away from each other. In other words, the modules establish an auxiliary channel for communication if they are not supervisor–subordinate, but need to share data. Therefore, the messages which will be passed using this channel need not be command or status messages. For this reason, the auxiliary messages are derived directly from NMLmsg and not from RCS_CMD_MSG or RCS_STAT_MSG. Similarly, the channel itself is defined as NML, as it is in Example 6.1, but not RCS_CMD_CHANNEL or RCS_STAT_CHANNEL.

As was discussed above, the subordinates of each module are assigned a unique identification number, returned by the function addSubordinate(). The purpose of the variable maneuver_sub_num is to hold the number of the maneuver module. If this module had other subordinates, there would be similar variables to hold their identification numbers also. These numbers will be used by the function sendCommand(), while sending commands to the particular subordinate, or accessing its status in the subordinate status array statusInData. However, for convenience you can define a variable for the status of each subordinate as part of the module and in a module-specific type in order to access the status of the given subordinate in a more clear and intuitive way. This is the reason for definition of the pointer maneuver_status that will later be assigned to point to statusInData[maneuver_sub_num].

Recall that the member functions of the class PRE_PROCESS, POST_PROCESS, and DECISION_PROCESS are also defined as virtual in the NML_MODULE base class. However, as we mentioned before, they are typically overloaded in each control module because the processing which will be done in these functions is application dependent. For example, the function DECISION_PROCESS simply calls one of the command functions based on the command message arrived; however, every module may accept different commands and therefore their function calls should be different. The command functions for the module above are INIT, HALT, and AUTODRIVE.

The variable supervisor_status is a pointer to the status message of this module (i.e., SUPERVISOR_STATUS) and will be passed as the last parameter to setStatChannel inside the constructor of the module. Then it will be sent to the NML channel every cycle for the superior of this module (e.g., the driver). Information or data which needs to be sent to the superior should be defined as an appropriate data member in SUPERVISOR_STATUS structure, and should be updated within the command functions of the module, or within the PRE_PROCESS or POST_PROCESS functions.

In Example 6.2 we present the constructor of the vehicle supervisor module, SUPERVISOR_MODULE, defined in Example 6.1. The example also contains the function INITIALIZE_NML(), which is used for the initialization of all the NML channels. Note, however, that the use of such a function is not mandatory and all the initializations done by this function can be done within the constructor. It has been defined here to show that the class SUPERVISOR_MODULE can be used as a base class for other modules. In that case, we may initialize all the NML channels within the derived module.

Example 6.2 also shows the use of some of the member functions of the NML_MODULE class discussed in the preceding section.

Example 6.2: Constructor for the SUPERVISOR_MODULE control module

```
// Constructor
SUPERVISOR_MODULE::SUPERVISOR_MODULE(int _is_base_class)
{
  if(!_is_base_class)
    {
      INITIALIZE_NML();
    }

  // Add additional code to initialize the module here.
  id = VEHICLE_ID; pltn_id = VEHICLE_ID;
  lane = 0; section = 0;
  optsize = 1; ownsize = 1; pos = 1;
  optspeed = 0; speed = 0;
  busy = 0;
}

void SUPERVISOR_MODULE::INITIALIZE_NML()
{
  setErrorLogChannel(new NML(nmlErrorFormat, "errlog",
                          "supervisor", "ivhs.nml"));
  setCmdChannel(new RCS_CMD_CHANNEL(supervisorFormat, "supervisor_cmd",
                                  "supervisor", "ivhs.nml"));

  supervisor_status = new SUPERVISOR_STATUS();
  setStatChannel(new RCS_STAT_CHANNEL(supervisorFormat, "supervisor_sts",
                               "supervisor", "ivhs.nml"), supervisor_status);

  maneuver_sub_num =
    addSubordinate(
   new RCS_CMD_CHANNEL(maneuverFormat, "maneuver_cmd",
                                    "supervisor", "ivhs.nml"),
   new  RCS_STAT_CHANNEL(maneuverFormat, "maneuver_sts",
                                      "supervisor", "ivhs.nml"));
  maneuver_status = (MANEUVER_STATUS *) statusInData[maneuver_sub_num];

  // auxiliary Input NML Channels
  //attn_to_sup
  ATTN_TO_SUP_CHANNEL = new NML(attn_to_supFormat, "attn_to_sup",
```

```
                                "supervisor", "ivhs.nml");
  attn_to_sup_data = (ATTN_TO_SUP_MSG *) ATTN_TO_SUP_CHANNEL->get_address();
}
```

The constructor here is used to set up the NML channels for the module as well as any other initialization that needs to be done. The command, status, and error log channels for this module are set up with the functions `setCmdChannel()`, `setStatChannel()`, and `setErrorLogChannel`. The connections to its subordinate are set up with `addSubordinate()`, whereas an auxiliary channel is set up simply by defining an NML communication channel. Note that the pointer variables, such as `maneuver_status` and `attn_to_sup_data`, which were defined for convenience, are initialized to point to the appropriate location of the memory. In other words, the variable `maneuver_status` is defined to point to `statusInData[maneuver_sub_num]`, where the status information of the maneuver module will be written, whereas the variable `attn_to_sup_data` is defined to point to the memory space where the auxiliary message, called `ATTN_TO_SUP_MSG`, from the attention module will be written. The address of this message is determined using the NML `get_address()` function. Finally, you can do any other needed initialization in the constructor for the module, such as initialization of any application-specific variables.

Recall from Figure 6.2 the operation of the `NML_MODULE`. The functions `PRE_PROCESS`, `DECISION_PROCESS`, and `POST_PROCESS` are called in this order. The function `DECISION_PROCESS`, on the other hand, calls one of the command functions of the module. Below we provide examples for all these functions and a simple command function.

The objective of the `PRE_PROCESS` function was to perform any operations that the module will perform at the beginning of each cycle, irrespective of the current command. Example 6.3 provides an example of a `PRE_PROCESS` function. It is a simple implementation of the `PRE_PROCESS` function of the drive module of the intelligent vehicle.

Example 6.3: Implementation of the `PRE_PROCESS` member function of the `DRIVE_MODULE`

```
/* PRE_PROCESS
   The PRE_PROCESS function is called every cycle after the command and
   subordinates status have been read but before DECISION_PROCESS is called.
   It is intended to be used for tasks such as sensory processing that should
   be performed every cycle regardless of the current command or state.
   */
void DRIVE_MODULE::PRE_PROCESS()
{
  // auxiliary Input NML Channels
  // Read new data from attn_to_drv
  ATTN_TO_DRV_CHANNEL->read();

  // Pre-Processing Code
  x_vel = attn_to_drv_data->x_vel;
```

```
    y_pos = attn_to_drv_data->y_pos;
    y_vel = attn_to_drv_data->y_vel;
    lane = attn_to_drv_data->lane;
    spacing = attn_to_drv_data->spacing;

    // Do any other processing
    // ...
}
```

Note from the example that each cycle the module reads the information from the auxiliary channel connecting it to the attention module (recall the `NML::read()` function discussed in Chapter 5). In the implementation of the real-world application this function will perform sensory processing, perhaps some conversions, and will read the information from the attention module. However, for simulation purposes it gets the data only from the attention module to use for the control algorithms (the variable `attn_to_drv_data` points to the data read from the auxiliary channel). If you do not want to update an auxiliary NML channel each cycle, then you may prefer not to call the `NML::read()` in the `PRE_PROCESS` function. Instead, you may prefer to read the information from the channel from somewhere within a command function only whenever necessary.

Note that each cycle the module performs similar read operations from its command channel and the status channels of its subordinates. However, these operations are transparent to the programmer since the module does them automatically (refer to Figure 6.2).

In Example 6.4 we provide an example `DECISION_PROCESS` function. It is the implementation of the `DECISION_PROCESS` function of the supervisor module.

Example 6.4: Implementation of the `DECISION_PROCESS` function of the `SUPERVISOR_MODULE`

```
/*
The DECISION_PROCESS function is called every cycle as long as there is a
non-zero command. It is expected to call a command function based on
commandInData->type.
*/
void SUPERVISOR_MODULE::DECISION_PROCESS()
{
  switch(commandInData->type)
    {
    case SUPERVISOR_INIT_TYPE:
      INIT((SUPERVISOR_INIT *)commandInData);
      break;

    case SUPERVISOR_HALT_TYPE:
      HALT((SUPERVISOR_HALT *)commandInData);
      break;

    case SUPERVISOR_AUTODRIVE_TYPE:
      AUTODRIVE((SUPERVISOR_AUTODRIVE *)commandInData);
      break;
```

```
    default:
      logError("The command %d is not recognized.",commandInData->type);
      break;
    }
}
```

The current command that is received by the given module is specified in the field `commandInData`. As you can see from above, the command function is called based on the value of this field. Every module may have different commands that it can execute. Most of the commands are application specific and are written by the RCS programmer. However, most of the modules will have a command for initializing the variables and parameters, as well as a command for stopping or halting the process. The `DECISION_PROCESS`, as in Example 6.4, is generated automatically by the RCS design tool. If you have more complicated logic for a module to decide on which operation to perform, you may implement this function in a different manner. Note, however, that you need to call one of the module's command functions within this function.

Example 6.5 provides a sample implementation of the `INIT` command performed by this module. You can develop the other command routines according to the objectives that should be achieved after executing a given command.

Example 6.5: Implementation of the `INIT` command function of the `SUPERVISOR_MODULE`

```
/* INIT
Parameter(s):
SUPERVISOR_INIT *cmd_in -- NML Message sent from superior.

Most Modules will have an INIT command.
The INIT function is expected to initialize any variables that may be
in an uninitialized or unknown state, send INIT commands to its subordinates,
wait for the subordinates to be DONE and then inform its superior that it is
done. The state tables should use the STATE_MATCH macro so the diagnostics
tool can highlight the current line in the state table.
*/
void SUPERVISOR_MODULE::INIT(SUPERVISOR_INIT *cmd_in)
{
  MANEUVER_INIT maneuverInitMsg;

  if(STATE_MATCH(NEW_COMMAND))
    {
      // Send an INIT command to all subordinates.
      sendCommand(&maneuverInitMsg, maneuver_sub_num);

      stateNext(S1);
      // Reinitialize variables here.

    }
  // Wait for all subordinates to report done.
  else if(STATE_MATCH(S1,
      maneuver_status->status == RCS_DONE &&
      1))
```

```
        {
          status = RCS_DONE;
          stateNext(S2);
        }
      else if(STATE_MATCH(S2))
        {
          // Idle State
        }
  }
```

The INIT function above provides a basic example for a state table of a command function. When a new command is received, the state is set to NEW_COMMAND. On the first cycle, the first if statement is true and the module sends an INIT command to its subordinate, the maneuver module, with the sendCommand function, sets the state to S1, and returns. In the next cycle the state table checks to see if the status of the maneuver module is RCS_DONE. If so, the module sets its own status to RCS_DONE and its state to S2, which is an idle state for this task. The states RCS_DONE, S1, and S2 are predefined. The user is not required to use these definitions in developing code. You can always define states based on the application; however, the state table should be implemented so that only one if statement is true at a time, which reduces the execution time in each cycle, and also allows the line of current execution to be tracked in the RCS diagnostics tool.

The implementation of the other command functions of the supervisor module and the other modules will be similar. However, they will follow the transition diagrams for the tasks presented in Chapter 4. We recommend that the reader try to implement the command functions of some of these transition diagrams as an exercise.

The objective of the POST_PROCESS function was to perform any operations that the module will perform at the end of each cycle, irrespective of the current command. Example 6.6 provides an example of a POST_PROCESS function. It is a simple implementation of the POST_PROCESS function of the attention module of the intelligent vehicle.

Example 6.6: Implementation of the POST_PROCESS function of the ATTENTION_MODULE

```
  /* POST_PROCESS
     The POST_PROCESS function is called every cycle after DECISION_PROCESS is
     called but before the status and the subordinates commands  have been
     written. It is intended to be used for tasks such as output filters that
     should be performed every cycle regardless of the current command or state.
  */
  void ATTENTION_MODULE::POST_PROCESS()
  {
    // Post-Processing Code

    // Update the auxiliary message to the drive module
```

```
    attn_to_drv_data.x_vel = x_vel;
    attn_to_drv_data.y_pos = y_pos;
    attn_to_drv_data.y_vel = y_vel;
    attn_to_drv_data.spacing = spacing;
    attn_to_drv_data.lane = lane;

    // Update the auxiliary message to the supervisor module
    // ...

    // Update the module's status message
    // ...

    // Do any other processing
    // ...

    // auxiliary Output NML Channels
    // Write data to attn_to_sup
    ATTN_TO_SUP_CHANNEL->write(&attn_to_sup_data);
    // Write data to attn_to_drv
    ATTN_TO_DRV_CHANNEL->write(&attn_to_drv_data);
}
```

In order for the drive and supervisor modules to read the correct information from the attention module, it needs to update the information to be passed and also to write this information to the channel. If you want these updates to be performed each cycle, the best place for that is in the POST_PROCESS function, as illustrated in Example 6.6 (recall the NML::write() function discussed in Chapter 5).

As in the case of the the PRE_PROCESS function, if you do not want to update an auxiliary NML channel each cycle, then you may prefer not to call the write in the POST_PROCESS function. Instead, you may call it from somewhere within a command function to update the information only whenever necessary.

Note again that in each cycle the module performs similar write operations to its status channel and the command channels of its subordinates. However, these operations are transparent to the programmer since the module does them automatically (refer to Figure 6.2).

In this section we discussed the derivation of a control module from the NML_MODULE as well as implementation of some of its function members. The most important functions are the command functions which implement the actual control algorithm for a given command or task request. Moreover, they can be also used for identification and estimation routines, planning and optimization, and so on. They implement the most application-dependent and difficult code of RCS programming. The programming of the rest is straightforward and easy. In fact, it is even possible to generate some of the code automatically using the RCS design tool, which we discuss in Chapter 10.

In the next section we discuss the implementation of the main loop of a control module via an example, and then conclude this chapter.

6.3 Writing the Main Loop(s)

In order to run the module it is necessary to call its `controller()` function from the main function for the program. In addition, the programmer has the option of combining several modules in the same executable, which eliminates some of the task switching and mutual exclusion overhead and allows for tighter synchronization. However, running the processes in a single executable makes the implementation of different cycle times more difficult. Therefore, we recommend that you run the processes with different timing as different executables. It is always a good idea to keep the code in the main function to a minimum so that the application can be reorganized easily.

Example 6.7 shows an example implementation of the main loop that creates and runs the supervisor and maneuver modules of the intelligent vehicle controller. We set up these two modules to run within the same executable with the same sampling time. The attention and drive modules run in another executable with a smaller sampling period than the supervisor and maneuver modules. This is because the sampling requirements of the algorithms on this level are more demanding. Note that it is not mandatory to run the modules within the same executable even if they have the same sampling time. However, putting them in the same executable may save switching time. On the other hand, if they are executing on different processors or backplanes, then they cannot be run within the same executable.

Example 6.7: Main function

```
/* ivhsmain.cc
   This file provides the C++ main function which
   creates and runs the following control modules:

   MANEUVER_MODULE
   SUPERVISOR_MODULE
   */

// Include Files
#include <stdlib.h>// exit()
#include <signal.h>// SIGINT, signal()
#include "rcs.hh"  // Common RCS definitions
#include "nml_mod.hh"  // NML_MODULE definitions
#include "maneuver.hh" // definition of MANEUVER_MODULE
#include "supervisor.hh" // definition of SUPERVISOR_MODULE

// flag signifying main loop is to terminate
int ivhs_done = 0;

//signal handler for ^C
extern "C" void ivhs_quit(int sig);
void ivhs_quit(int sig)
{
  ivhs_done = 1;
}

// main loop, running 2 controller(s)
```

```
int main(int argc, char **argv)
{
  set_rcs_print_destination(RCS_PRINT_TO_STDOUT);

  RCS_TIMER *timer = new RCS_TIMER(0.1);
  MANEUVER_MODULE *maneuver = new MANEUVER_MODULE();
  SUPERVISOR_MODULE *supervisor = new SUPERVISOR_MODULE();

  // set the SIGINT handler
  signal(SIGINT, ivhs_quit);

  // enter main loop
  while(!ivhs_done)
    {
      maneuver->controller();
      supervisor->controller();

      timer->wait();
    }

  // Delete Modules
  delete maneuver;
  delete supervisor;

  // Delete Timer
  delete timer;
}
```

Note that the file includes the header files containing the definitions of the module classes that it will run and creates an object from each class using the operator `new`. Moreover, a timer is created with the required sampling period (which is 0.1 in the example above). Inside a `while` loop, the `controller` function for each module is called, followed by a wait function that puts the processes to sleep until time for the next cycle. The `RCS_TIMER::wait` function is specifically designed for this, and sleeps for a period that varies depending on the amount of processing the modules do, so that the period between cycles remains consistent. For example, suppose that for one cycle the sum of the time required to run all of the controllers is 40 ms and the timer was initialized at 0.1 s = 100 ms. The `wait` function will put the process to sleep 60 ms so the total time for the cycle is 100 ms. (See Chapter 8 for more detailed information about the RCS timer.)

Sometimes, it is necessary to have some way of breaking the loop to shut down the controller gracefully. In this example a signal handler for `"<CNTRL>-C"` is set up to do this through the function `signal`. Whenever `"<CNTRL>-C"` is pressed, the function `ivhs_quit`, passed as an argument to `signal`, is called which assigns 1 to the variable `ivhs_done`, and this breaks the loop.

In this chapter we discussed the `NML_MODULE` base class. We showed how you can derive a controller module from this class and how to implement some of the member functions which are application specific. Moreover, we showed how the modules can be created and called within the main loop of the program. These parts will be most useful for the beginner RCS programmer and will be used mostly during the development of the control software for the application in the succeeding chapters.

Before moving on to the discussion of the NML configuration files we want to emphasize that almost all of the code discussed in this chapter can be generated automatically using the RCS design tool, which is a Java-based graphical tool for RCS application development that is discussed in Chapter 10. The discussion here was presented for the sake of completeness and so that the reader gets a deeper understanding of the operation and use of the `NML_MODULE`.

Chapter 7

Writing NML Configuration Files

NML configuration files are *ascii* files which should be written by the NML application programmer. All the buffers and processes in the application, as well as their locations, are defined in this file. The aim of a configuration file is basically to allow the user to change the configuration of CMS and NML without recompiling and relinking the code for the whole program. Therefore, all the options available in CMS can be specified in the configuration file. These include the communication protocols and means for mutual exclusion of shared memory. We recommend that the reader who is not familiar with these concepts consult Appendix C for a brief introduction, or one of the many books in the area, such as [11, 50] on communication networks and [15, 37] on operating systems, for a detailed analysis.

Before proceeding, recall how to establish NML communication channels. We repeat here the format of the NML constructor for convenience of the reader:

```
NML(NML_FORMAT_PTR f_ptr, char *buf, char *proc, char *file);
```

The last parameter, *file*, that we provide to the constructor is the name of the NML configuration file. Moreover, the parameter *buf* is the name of the NML shared memory buffer to connect to, and the parameter *proc* is the process name under which to access the buffer. The rest of the configuration parameters for the NML communication channel will be read from the corresponding lines for the corresponding buffers and processes within the configuration file.

In an NML configuration file, lines beginning with different characters supply different information. The lines beginning with # or spaces are *comment lines*, those beginning with B specify information about a particular buffer and are called *buffer lines*, and others which begin with P specify information about how a particular process accesses a buffer and are called *process lines*. We begin with description of basic buffer and process types available.

7.1 Process and Buffer Types

Every buffer line contains information about the type of the buffer defined, that is, how the processes in the application will connect to this buffer. According to the relative location of a given process to this buffer it can be defined as local or remote. Local processes are those which run on the same computer with the buffer and they are described with the keyword `LOCAL` on the process line. Remote processes are the processes which run on a different computer or backplane and they are described with the keyword `REMOTE`.

Another option for describing a process is to use the keyword `PHANTOM`. This option is used whenever the user wants to redirect temporarily the messages for a CMS buffer (see the corresponding part of Chapter 5 for more information on phantom processes and buffers). Local processes can connect to buffers using `GLOBMEM, SHMEM, FILEMEM`, or `LOCMEM`. Remote processes can connect using `TCP, UDP`, or `RPC`.

GLOBMEM: `GLOBMEM` is intended for communication across a VME backplane between processors with minimal operating system intervention, which allows greater portability and interoperability than `SHMEM`, but users need to specify the addresses of memory to use, and mutual exclusion is implemented with a set of locks in memory rather than with semaphores provided by the operating system. `GLOBMEM` may also be used with a Bit3 ISA or EISA to a VME adapter.

SHMEM: `SHMEM` is intended for communications between tasks managed by the same operating system. The operating system allocates the memory to be shared so users do not need to specify one. Users have a choice of mutual exclusion techniques which include something similar to `GLOBMEM`'s method, using an operating system semaphore or mutex, or disabling and enabling context switching or interrupts during the appropriate critical sections.

One of the most significant ways of optimizing shared memory accesses is to choose the fastest form of mutual exclusion that still allows the application to run reliably. The type of mutual exclusion can be selected by placing `mutex=option` after `SHMEM` keyword. The available options are described next.

- `mutex=os_sem`
  This is the default. An operating system semaphore is created using the same key as that used for shared memory. This semaphore is taken before each read or write and released afterward. If another process already has taken, but not released, the semaphore when the semaphore is taken, the operating system puts the process to sleep until the semaphore is released. This is reliable; however, the operating system calls to take and release the semaphore take from 100 to 500 microseconds to execute, depending on the CPU and operating system. This is a very long time compared to the time required to copy small messages (<10k) into the shared memory area.

- `mutex=none`
  The library will make no attempt to provide mutual exclusion. This is generally a dangerous option that is recommended only when some other form of mutual exclusion is implemented on top of NML. As the reader will see in Chapter 8, the RCS library has some tools for mutual exclusion. The programmer may want to implement an application-specific type of mutual exclusion using the RCS tools or system tools. In that case this option can safely be chosen.

- `mutex=no_interrupts`
  If there is only a single processor running multiple tasks or threads, most operating systems will be prevented from switching tasks, or threads, as long as interrupts are disabled. Since disabling and enabling interrupts is often much faster than the semaphore calls, this can be an efficient means of mutual exclusion. Currently, this option will work only under Lynx and VxWorks. Under Lynx it is necessary to run the process as root. You also need to know how long your system can tolerate having interrupts disabled.

- `mutex=no_switching`
  Some operating systems allow you to disable task switching, which can be used just as disabling interrupts above, except that it is a little gentler in that it prevents other tasks from executing but allows hardware interrupts to execute. Currently, this option will only work under Lynx and VxWorks. Under Lynx it is necessary to run the process as root.

- `mutex=mao split`
  This is an experimental option. It provides mutual exclusion by splitting the buffer in half. At any given time one half is used for reading, while the other half is used for writing. Only at the end of a write is a flag changed that switches which half is used for what. A small area at the beginning of the buffer is used for registering which processes are reading and which are writing at any given time. This allows many readers and one writer to access the buffer simultaneously without allowing any reader to get a partial message or require semaphore calls. However, situations can occur where it is necessary to force one of the processes to wait, or return an error. When a process is forced to wait, there is currently no provision for priority inheritance, which means that currently while most of the time this option is very fast compared to the others, it can occasionally take significantly longer, or timeout.

FILEMEM: `FILEMEM` is useful primarily for debugging. Messages are read from and written to a file, or messages may be read from one file and written to another. Record locking is used to provide mutual exclusion. This is the least efficient method; however, the output file can be used for postmortem analysis, and scripts can be written and used as the input file for unit testing. `FILEMEM` buffers must have *neutral* set to 1 and add `disp` to the buffer line to use the displayable method of neutral data encoding.

LOCMEM: `LOCMEM` is useful when many modules are linked together in one thread of execution but you want to write them so that each module uses the NML API (application program interface) just as it would if it needed to communicate with the other modules running separately. There is no need for any mutual exclusion mechanism, and memory is obtained with a simple malloc so the operating system will not exceed its limits for semaphores or shared memory segments.

RPC: `RPC` stands for *remote procedure call.* There are two incompatible types of RPC, although they are both intended for the same purpose. CMS uses *open network computing* (ONC) RPC, which is available on more of the platforms we were interested in rather than the *network computing system* (NCS) RPC which was chosen by the *Open Software Foundation* (OSF) and is distributed with Visual C++. RPC was the first remote connection method available to NML users. Originally, RPC was required for its dispatch services and eXternal Data Representation (XDR) interface, but these features are now provided in CMS on top of TCP or UDP. NML users that select RPC will not need to use any of the RPC API (just as they would not use any of the socket APIs if TCP or UDP were selected); however, they can check to see that the servers are registered with the UNIX command `rpcinfo`.

TCP: This is the recommended method for remote connection for most applications. It is more reliable and can handle larger messages than UDP, and is moreover, more widely available than RPC.

STCP: `STCP` is a simplified form of the software on top of TCP. It is useful as an interface to languages other than C++/Java, where creating a binary version of the NML message structure(s) is difficult and/or there is no version of the RCS library available. It allows applications in that language to send and receive simple text strings with low-level socket functions, while C++ and Java applications still use the normal NML API. It must be used with the display neutral encoding method. This means that `disp` must be placed on the buffer line, and `xdr` or `ascii` must be removed if present.

UDP: UDP is not as reliable as TCP. However, it is faster, and therefore it can be used in applications which require very fast communication. Note that if UDP is used, messages should not exceed 512 bytes.

The interested reader may consult Appendix C or [11, 50] for more information on network communication protocols.

7.2 Buffer Lines

The format of buffer lines is

```
B name type host size neut RPC# buffer# max_procs [Type-spec data]
```

- **B** is the literal character identifying the line as describing a buffer.

- **name** is a string identifying the buffer. The programmer can use any valid identifier as a buffer name. In the program you should run NML servers for the buffer with the name defined in the configuration file. Moreover, the buffer name passed to the NML constructor should match the name of the buffer defined here.

- **type** is a string which can currently be `SHMEM` for shared memory buffers used for communications between processes on the same CPU, `GLOBMEM` for global memory buffers on a VME or EISA backplane, `LOCMEM` for a communications between functions run within the same process (multi-threaded applications should never use `LOCMEM`), or `PHANTOM` for buffers that use phantom functions.

- **host** is the Internet name of the host for the buffer. If remote processes need to access the buffer, they will look for an NML server on this host.

- **size** is the byte size of the buffer in decimal.

- **neut** should be either 0 or 1. The number 1 indicates that the data stored in the buffer should be formatted into a machine-independent or neutral format. Remote processes can still access buffers with a 0 even when running on a processor using incompatible types because the NML server can encode and decode automatically. However, the buffer itself must be encoded (neut=1) if the buffer will be accessed locally by processors of incompatible types.

- **RPC#** represents a remote procedure call number. All buffers that will be accessed using the same server should have the same RPC#. RPC program numbers 0x20000000 to 0x30000000 are for users to define. You may use the `rpcinfo` command on UNIX workstations to determine which numbers are already being used. Currently, the RPC# here is still required as a placeholder even if TCP or UDP were to be used instead. This allows for some backward compatibility with configuration files written before TCP or UDP were options; however, this backward compatibility may be dropped in the future.

- **buffer#** A server for more than one buffer will use the buffer number to determine which buffer to access for each request. Each buffer should be given a unique buffer number.

- **max_procs** represents the maximum number of processes allowed to use this buffer. (If the total possible connections equals n, then the connection number in the process lines may be 0 to $n-1$.)

- The area following the `max_procs` number is used to store information relevant to particular types of buffers.

`GLOBMEM` buffers require a physical address. The physical address may be specified with one or more address equations of the form `???_addr=0xnnnnnnnn` where `???` may be a host name, process name, or bus name.

`SHMEM` buffers expect an integer key in the type-specific area. The key needs to be a unique shared memory key on your system. For many UNIX systems the command `ipcs` can tell you which shared memory keys are in use. All of the NML buffers using `SHMEM` on a particular host need to have different keys.

Configuring the file for use of TCP or UDP instead of RPC for remote access to the buffer is simple—you can add either `TCP=` or `UDP=` and a port number to the buffer line. It is required to use a port number greater than 1024 since ports 1–1023 are reserved for processes with user identification 0. It is recommended that the programmer selects a port between 5000 and 32000 to minimize the chance of conflicting with another application.

To use an alternative neutral encoding scheme where the messages are converted to *ascii* or plain text strings, you should add the word `ascii` to the buffer line, or `disp` to use a neutral encoding scheme that produces text strings which look much better for display. (The `disp` is required for `FILEMEM`.) Another option would be to add `xdr` if you want the data stored in the buffer to be in XDR format.

To set the input file for `FILEMEM`, you can add `in=` and the filename to the buffer line. Without this, `FILEMEM` will use `stdin`. Similarly, it is possible to set the output file for `FILEMEM` by adding `out=` and the filename to the buffer line. If the filename is omitted, `FILEMEM` will use `stdout`. It is possible to force `FILEMEM` to limit the number of messages in the output file by adding `max_out=` and the maximum number of messages to store in the output file. When this limit is reached, the file will be overwritten starting at the beginning. It is possible to enable queuing of messages in the buffer. To do this you should add the word `queue` to the buffer line. Finally, if needed, one can set up a password to the buffer to prevent or restrict access of some users to that buffer. This is done by adding `passwd=file_name.pwd` to the end of each buffer line and then using the program `nmlpwd` to create or modify the `.pwd` file.

In Example 7.1 we provide an example of buffer definitions within an NML configuration file. Most of these buffers were used in the examples in Chapter 5.

Example 7.1: Definition of NML buffers

```
# ex_cfg.nml
# buffers:
# Name       Type    Host   size neut? RPC#    buf#  MP [type-spec]
#           GLOBMEM host   size neut  RPC#    buf#  MP phys_addr
```

```
#           SHMEM    host    size neut  RPC#        buf#  MP key
B ex_buf1   SHMEM    dopey   512    0 0x20001050    0     10 101 UDP=5001 ascii queue
B ex_buf2   SHMEM    dopey   512    0 0x20001050    1      3 102 TCP=6001 ascii
B ex_buf3   SHMEM    dopey   512    0 0x20001051    2      1 103 STCP=9001 ascii
B ex_buf4   GLOBMEM vx40     512    0 0x20001060    3     11 vme_addr=4e00000
```

In this example, the first three buffers are defined as SHMEM, they are encoded in ascii format, and are situated on the host called dopey. The last buffer is of type GLOBMEM and it is on the host called vx40 on the address 4e00000. Note that while SHMEM buffers have some integer key specified after the maximum process number (101, 102, and 103 in the file), the GLOBMEM does not. These numbers need to be unique shared memory keys for the system. Therefore, you have to change them if there are already other shared memory segments using these numbers. As mentioned before, on a UNIX system the command ipcs can be used to find which shared memory keys are already in use. All the buffers are of size 512 bytes, and they are not encoded because they have 0 in the neut field. The first buffer is accessed by 10 processes, the second one by 3 processes, the third buffer by 1 process, and the last one by 11 processes. The communication protocols, to be used by remote processes to access the shared memory buffers, are defined UDP on port 5001 for ex_buf1, TCP on port 6001 for ex_buf2, and STCP on port 9001 for ex_buf3. The buffer called ex_buf1 is defined as a queue, which means that the messages will be queued in the buffer, i.e., new coming messages will not overwrite the old ones.

Every RCS implementation should have its specific structure. The application programmer decides during the task analysis, and controller hierarchy definition for the application, how many buffers are needed and on which host they should be located for better performance. Moreover, the user has a choice of how to perform mutual exclusion for the buffers and communication protocols for remote processes connecting to the buffer. In general, it is safe to choose the default mutual exclusion, and you can choose the communication protocol either as TCP or UDP, based on the reliability and speed requirements of the system. The size of the buffers depends on the size of the information that will be stored in them—in other words, the size of the NML messages that we will need to pass from one process to another.

7.3 Process Lines

The format of process lines is

```
P  name  buffer  type  host  ops  server  timeout  master  c_num
```

- **P** is the literal character identifying the line as describing a process.
- **name** is the process name which should match the process name passed to the CMS config function or the NML constructor.

- **buffer** is the string matching the identifier used in a buffer line to which this process intends to connect.

- **type** is a string which describes the location of the process relative to the corresponding buffer. It can currently be `LOCAL`, `REMOTE`, or `PHANTOM`.

- **host** is the Internet name of the computer that the process is expected to run on.

- **ops** is a string indicating whether the process will read or write. It can be `R` for readers, `W` for writers, or `RW` for processes which will both read and write.

- **server** should be either 0, 1, or 2. A 1 indicates that this process is responsible for running a server for the buffer, a 2 indicates that the `nml_start` function (see Chapter 5 for information on `nml_start` function) should spawn a server for the buffer, and a 0 is for processes that will neither spawn nor run a server for the buffer.

- **timeout** is the time in seconds for the process to wait for the mutual exclusion semaphore or for an RPC call to return before an error is reported. A time can be specified in floating point and the timeout is measured as close as possible for the given platform and process type, or `INF` to never timeout.

- **master** can be either 0 or 1. A 1 indicates that the process is a master for this buffer. A master process is responsible for creating the buffer, i.e., initializing a memory area for it. Therefore, every buffer should have, and can have, only one master.

- **c_num** is the connection number for this process to the buffer specified in the line. It must be between 0 and $n-1$, where n is the number of total connections specified on the buffer line of that buffer. Different processes connect to the same buffer with different connection numbers.

Now we provide Example 7.2, to illustrate the definition of the process lines in a configuration file. Some of these processes were used in the examples in Chapter 5; others are defined in order to provide to the reader a sample definition. These processes use the buffers defined in Example 7.1.

Example 7.2: Definition of processes

```
# ex_cfg.nml
# processes:
# name          buffer   type    host    ops   server  timeout master c_num
P ex1_proc        ex_buf1 LOCAL   dopey   W       0        INF      1     0
P ex2_proc        ex_buf1 LOCAL   dopey   W       0        INF      0     1
P ex2_proc        ex_buf2 LOCAL   dopey   W       0        INF      1     0
P ex3_proc        ex_buf1 LOCAL   dopey   R       0        INF      0     2
P ex4_proc        ex_buf1 LOCAL   dopey   W       0        INF      0     3
P ex5_proc        ex_buf1 PHANTOM dopey   W       0        0.5      0     4
```

```
P ex6_proc      ex_buf1 LOCAL   dopey   W     0     0.5    0    5
P ex8_proc      ex_buf1 REMOTE  rosie   RW    0     10.0   0    6
P ex8_svr       ex_buf1 LOCAL   dopey   RW    1     5.0    0    7
P ex8_svr       ex_buf3 LOCAL   dopey   RW    1     5.0    1    0
P ex9_svr       ex_buf1 LOCAL   dopey   RW    1     5.0    0    8
P ex9_svr       ex_buf2 LOCAL   dopey   RW    1     5.0    0    1
P ex10_svr      ex_buf1 LOCAL   dopey   RW    1     5.0    0    9
P ex10_svr      ex_buf2 LOCAL   dopey   RW    1     5.0    0    2
```

Most of the processes defined above will run on the computer called `dopey`; therefore, they are defined as `LOCAL` to the buffers which are located on the same host. The procedure `ex5_proc` is an example of a `PHANTOM` process and the process `ex8_proc` is located on host `rosie` and therefore is defined as `REMOTE`. Whether a process is a reader, writer, or both to a buffer is indicated with the letters `R`, `W`, or `RW`, respectively. For instance, `ex1_proc` writes to `ex_buf1`, `ex3_proc` reads it, whereas `ex8_proc` both reads it and writes to it. In general, the processes configured as servers are configured to both read from and write to a buffer since they access the buffer for all remote processes using it, and some of them may be reading it and some writing to it. The last processes with 1 in the `server` field act as servers for the buffers in the corresponding line, and the timeout for reporting an error is as specified in the `timeout` field. Therefore, `ex1_proc` will wait for the semaphore of buffer `ex_buf1` to be released indefinitely, while the procedure `ex10_svr` will wait just 5.0 seconds and then report an error. Every buffer can have only one master, which is indicated with 1 in the `master` field in the configuration file. The master process for a buffer is responsible for creating the buffer, therefore it is required to be run first. The master for `ex_buf2` is `ex2_proc`; therefore, it must be run before `ex9_svr` and `ex10_svr`. Every process is using a unique connection number to connect to a given buffer. For example, `ex2_proc`, `ex9_svr`, and `ex10_svr` connect to the buffer `ex_buf2` using 0, 1, and 2, respectively, as connection numbers.

It is a good practice to make the NML server master for the buffer it is writing on. This allows the NML server to be run first and communications get ready before running actual controllers. You can equally well define the process to be the master for the buffer it is accessing; however, this will require the process to be run first. This is because the buffer needs to be created before any other process can connect to it and the master is responsible for creating the buffer.

It is possible to use different communication protocols simultaneously for the same buffer. In that case, you can copy the configuration file to a different directory and modify it so that a different protocol is used; for example, replace TCP with UDP and also the port numbers. Another required change is that in the new file there should not be masters defined because there is already one process set up as a master. Therefore, all the 1's in the `master` field should be set to 0's. Then it is enough to run the NML servers from the directory with the new copy of the configuration file.

We conclude this section with a candidate configuration file for the intelligent vehicle's controller that is shown in Example 7.3. This file assumes that all the buffers are located on the same computer, called `eepc172` (the computer on the vehicle board or in our case, the computer on which the program for this particular vehicle is running).

Example 7.3: NML configuration file for the intelligent vehicle

```
# NML Configuration file for the ivhs application

# Buffers
# Name              Type  Host    size neut? RPC#      buf# MP  . . .
B attention_cmd  SHMEM eepc172 2048  0  0x2089b28d  1  3 64080 TCP=7893 xdr
B attention_sts  SHMEM eepc172 2048  0  0x2089b28d  2  3 64081 TCP=7893 xdr
B drive_cmd      SHMEM eepc172 2048  0  0x2089b28d  3  3 64082 TCP=7893 xdr
B drive_sts      SHMEM eepc172 2048  0  0x2089b28d  4  3 64083 TCP=7893 xdr
B maneuver_cmd   SHMEM eepc172 2048  0  0x2089b28d  5  3 64084 TCP=7893 xdr
B maneuver_sts   SHMEM eepc172 2048  0  0x2089b28d  6  3 64085 TCP=7893 xdr
B supervisor_cmd SHMEM eepc172 2048  0  0x2089b28d  7  2 64086 TCP=7893 xdr
B supervisor_sts SHMEM eepc172 2048  0  0x2089b28d  8  2 64087 TCP=7893 xdr
B errlog         SHMEM eepc172 8192  0  0x2089b28d  9  5 64088 TCP=7893 xdr queue
B attn_to_sup    SHMEM eepc172 2048  0  0x2089b28d 10  3 64110 TCP=7893 xdr
B attn_to_drv    SHMEM eepc172 2048  0  0x2089b28d 11  3 64111 TCP=7893 xdr

# Processes
# Name          Buffer          Type    Host    Ops server? timeout master? cnum

# attention(0)
P attention    attention_cmd  LOCAL  eepc172  R    0       0.1      0       0
P attention    attention_sts  LOCAL  eepc172  W    0       0.1      0       0
P attention    attn_to_sup    LOCAL  eepc172  W    0       0.1      0       0
P attention    attn_to_drv    LOCAL  eepc172  W    0       0.1      0       0
P attention    errlog         LOCAL  eepc172  W    0       0.1      0       0

# drive(1)
P drive        drive_cmd      LOCAL  eepc172  R    0        0.1     0       0
P drive        drive_sts      LOCAL  eepc172  W    0        0.1     0       0
P drive        attn_to_drv    LOCAL  eepc172  R    0        0.1     0       1
P drive        errlog         LOCAL  eepc172  W    0        0.1     0       1

# maneuver(2)
P maneuver     maneuver_cmd   LOCAL  eepc172  R    0        0.1     0       0
P maneuver     maneuver_sts   LOCAL  eepc172  W    0        0.1     0       0
P maneuver     attention_cmd  LOCAL  eepc172  W    0        0.1     0       1
P maneuver     attention_sts  LOCAL  eepc172  R    0        0.1     0       1
P maneuver     drive_cmd      LOCAL  eepc172  W    0        0.1     0       1
P maneuver     drive_sts      LOCAL  eepc172  R    0        0.1     0       1
P maneuver     errlog         LOCAL  eepc172  W    0        0.1     0       2

# supervisor(3)
P supervisor   supervisor_cmd LOCAL  eepc172  R    0        0.1     0       0
P supervisor   supervisor_sts LOCAL  eepc172  W    0        0.1     0       0
P supervisor   maneuver_cmd   LOCAL  eepc172  W    0        0.1     0       1
P supervisor   maneuver_sts   LOCAL  eepc172  R    0        0.1     0       1
P supervisor   attn_to_sup    LOCAL  eepc172  R    0        0.1     0       1
P supervisor   errlog         LOCAL  eepc172  W    0        0.1     0       3
```

```
# ivhssvr(4)
P ivhssvr        attention_cmd  LOCAL  eepc172 RW    1       0.1    1     2
P ivhssvr        attention_sts  LOCAL  eepc172 RW    1       0.1    1     2
P ivhssvr        attn_to_drv    LOCAL  eepc172 RW    1       0.1    1     2
P ivhssvr        attn_to_sup    LOCAL  eepc172 RW    1       0.1    1     2
P ivhssvr        drive_cmd      LOCAL  eepc172 RW    1       0.1    1     2
P ivhssvr        drive_sts      LOCAL  eepc172 RW    1       0.1    1     2
P ivhssvr        errlog         LOCAL  eepc172 RW    1       0.1    1     4
P ivhssvr        maneuver_cmd   LOCAL  eepc172 RW    1       0.1    1     2
P ivhssvr        maneuver_sts   LOCAL  eepc172 RW    1       0.1    1     2
P ivhssvr        supervisor_cmd LOCAL  eepc172 RW    1       0.1    1     1
P ivhssvr        supervisor_sts LOCAL  eepc172 RW    1       0.1    1     1
```

Note that the buffers and processes in the configuration file correspond to those discussed in the examples in Chapters 5 and 6. In other words, in the file above we have the definition of all the command, status, and auxiliary buffers for the intelligent vehicle of our design problem.

7.4 Version 2 NML Configuration File Format and Tool

The NML configuration file format described so far is called the version 1 configuration file format. It has remained stable since 1995 except for some rarely used options that can be placed at the end of buffer and process lines. That format is also the only format that NML constructors can currently use directly, and it is the format generated by the NML configuration file builder (a CGI script and HTML form) and by the RCS design tool.

In this section we describe a configuration format that was proposed later (version 2). The goals for this new configuration format are:

- To make it easier for multiple organizations to share configuration files.
- To allow less verbose files by allowing users to specify only items that need to be different than the default.
- To reduce the number of errors caused by novice users.
- To eliminate some tedious tasks.

The version 2 configuration format is supported only through a tool that takes files written in the new format and produces files in the older format, described earlier in this chapter. These files can then be used in an application in the same way as files written by hand in the older format or generated by one of the graphical tools.

7.4.1 Configuration File Line Types

The new configuration files are once again *ascii* text files that contain eight different types of lines that can be identified by the first character or word on the line as shown in Table 7.1.

Table 7.1: Line types for the version 2 NML configuration files.

Starting Word or Character	Line Description
`#`	Specifies comment lines. They have no effect on the output file.
`##`	Describes insert lines. They have the same effect as regular comment lines except that they are inserted as comments in the output file.
`include`	Include lines cause the contents of another file to be read as if the text were included at that point in the original file.
`define`	Definition lines define a variable that can be used through the rest of the file, by preceding the variable name with "`$`" and surrounding it with parentheses, like this: "`$(varname)`."
`buffer_default`	Buffer default lines set defaults that affect buffer lines that occur after this line.
`b`	Specifies buffer lines. Each buffer line must contain "`name=`" and the name of the buffer somewhere on the line. These lines are used to create an entry for a particular buffer.
`process_default`	Process default lines set defaults that affect process lines that occur after this line.
`p`	Describes process lines. Each process line must contain either "`name=`" or "`bufname=`" somewhere on the line and are used to create an entry linking a particular process to a particular buffer. For proper checking, process lines must occur after the buffer line for the buffer to which the process is connecting.

7.4.2 Buffer Variables

The following variables, with the exception of "`name`," can be set either on a buffer line to modify only one buffer or on a buffer default line to affect several buffers. Note that these are the same buffer variables described earlier. We repeat them here with some additional information.

- `name` – The name of the buffer that is passed to the NML constructor. It must be unique. There is no default value; therefore, this variable must be set on each line that begins with "`b`" and cannot be used in `buffer_default` lines.

- `buftype` – The type of buffer. It can be `shmem`, `globmem`, `filemem`, `locmem`, or `phantom`. The default value is `shmem`.

- `host` – The host name where a server must be run if any processes are going to connect remotely. The default value is `localhost`.

- `size` – The size of the largest message that can be sent to the buffer. The amount of memory allocated will be slightly larger than this to accommodate some handshaking flags. The default value is 960.

- `neutral` – Whether the local buffer should be neutrally encoded. Buffers should be neutrally encoded if they can be accessed by multiple CPU types by a Bit3 adaptor for example or to force messages to go through format and update functions that significantly reduces message size, such as when variable-length arrays are used. Messages are always neutrally encoded when sent over a network. The default value is `false`.

- `bufnumber` – A unique number used to identify the buffer within a server. The default value is calculated based on the position of the buffer line in the file. If its default value is modified with a default buffer line, the default value for subsequent lines will still be incremented from this starting value.

- `max_proc` – The maximum number of processes that can connect to a buffer locally. It does not affect remote processes or `shmem` buffers using the default mutual exclusion mechanism. The default value is calculated based on the number of processes connecting to this buffer.

- `key` – A unique number used to identify the shared memory and semaphore used for mutual exclusion in a shared memory buffer. It is relevant when using the `ipcs` or `ipcrm` commands. The default value is calculated based on the position of the buffer line in the file.

- `bsem` – A unique number used to identify the semaphore used for blocking reads. The default value is -1, so blocking reads are not allowed by default. However, if the value is changed with a default buffer line, then subsequent lines will increment this starting value.

- `vme_addr` – The VME address used for GLOBMEM on a VME backplane. The default value is 0, which is unusable. However, if the default value is changed with a default buffer line, then subsequent buffers will use the sum of this value plus the size of the preceding buffers.

- `remotetype` – The protocol that should be used by remote processes connecting to this buffer which could be `tcp`, `stcp`, or `udp`. The default value is `tcp`.

- `port` – The TCP or UDP port used by remote processes. The default value is 30000.

- `enc` – The nuetral encoding method, which can be `xdr`, `ascii`, or `disp`. The default value is `xdr`.

- `queue` – Whether messages in this buffer should be queued. Setting it to a value greater than 1 also multiplies the size of the buffer by this value. The default value is 0.

- `diag` – Whether to enable supplemental timing diagnostic information to be logged to the buffer.

Note that some of the flags that were available in the old configuration file were not described here. In other words, the new tool will not recognize them. This will produce a warning, but the old flag should simply be pasted to the end of the buffer line generated, which should allow use of the unrecognized flag.

7.4.3 Process Variables

The following variables can be placed either on a process line to affect only one process connection to one buffer or on a default process line to affect multiple connections. Note that these are the same process variables described earlier. We repeat them here with some additional information.

- `name` – The name of the process that is passed to the NML constructor. There is no default value; this variable must be set by the user.

- `bufname` – The name of the buffer to which this process is connecting. There is no default value; this variable must be set by the user.

- `proctype` – The type of process, which can be `local` or `remote`. The default is `remote`.

- `host` – The name of the host, in which this process is running. It is currently used only to add comments to the output file. The default value is `localhost`.

- `ops` – The operations allowed by this process on this buffer, which can be `r`, `w`, or `rw` (for `READ_ONLY`, `WRITE_ONLY`, and `READ_WRITE`, respectively). The default value is `rw`.

- `timeout` – The time in seconds to allow before this process should timeout waiting for a read or write, specified as a double or `INF` to indicate infinity (i.e., no timeout). The default value is `INF`.

- `master` – Whether this process will be the master of this buffer. The master creates and clears the buffer when it is started. The default value is `false`.

- `server` – Whether this process will act as a server for this buffer. The value of 2 has the special meaning that the process will spawn a server but then continue to access the buffer locally. The default value is `false` or 0.
- `c_num` – The connection number, which should be unique among processes connecting to the same buffer locally. The default value is calculated based on the number of processes that connected to this buffer before.
- `sub` – The subscription interval in seconds for remote processes. The default value is -1, which indicates no subscription.

7.4.4 Tool Command Line Arguments

As mentioned before, NML constructors cannot currently read the new NML configuration file format. Therefore, there is a need for a tool to convert the files from the new format to the old format. The program `nmlcfg` can be used for this purpose.

Use of the `nmlcfg` command is as

```
nmlcfg [-D name=val] [-I dir] [-o output_file] files . . .
```

Here the meanings of the flags are as follows:

- `-D name=value` – This flag defines a value on the command line just as if a `define` line were used.
- `-I directory` – This flag adds a directory to a list of directories to be searched if an `include` line is found.
- `-o output_file` – This flag is to specify the file where the output file (i.e., the old format NML file) will be written.

All command line arguments that are not preceded by one of the flags above are considered to be input files that should be in the new format. If more than one input file is specified, then they are read in the order they occur and only one output file will be produced. If no output file is specified, the data will be printed on the screen.

7.4.5 Examples

Now we provide an illustrative example of a version 2 NML configuration file. Assume that we have created a file called "myhosts" that contains the following:

```
# Set host aliases.
define host1=dopey
```

This file could be used to change the host names for the buffers and processes.

Example 7.4 shows a sample version 2 NML configuration file that creates two buffers and three processes. Both of the buffers are placed on `host1`, which is defined to be dopey in the file "myhosts" that is included by our NML file. The first process, named mysvr, is defined to be a server, master, and local to all the buffers that it accesses, whereas the other two processes, named proc1 and proc2, are defined not to be servers and masters while preserving the other process default values.

Example 7.4: Example version 2.0 NML configuration file

```
include myhosts

buffer_default host=$(host1)
b name=ex_cmd
b name=ex_stat

process_default server=1 master=1 proctype=local name=mysvr
p bufname=ex_cmd
p bufname=ex_stat

process_default server=0 master=0

process_default name=proc1
p bufname=ex_cmd
p bufname=ex_stat


process_default name=proc2
p bufname=ex_cmd
p bufname=ex_stat
```

The old-style configuration file generated from the file in Example 7.4 is shown in Example 7.5.

Example 7.5: Old-style configuration file corresponding to the NML file in Example 7.4

```
# Buffers:
#   name    type    host   size  neut  0  buf#  max_proc  . . .
B  ex_cmd  SHMEM   dopey  1024   0    0   1       4      10001  TCP=30000  xdr
B  ex_stat SHMEM   dopey  1024   0    0   2       4      10002  TCP=30000  xdr


# Processes:
# Name  Buffer  type    host       ops  server  timeout  master c_num  . . .
P mysvr ex_cmd  LOCAL localhost   RW     1       INF      1       0
P mysvr ex_stat LOCAL localhost   RW     1       INF      1       0
P proc1 ex_cmd  LOCAL localhost   RW     0       INF      0       1
P proc1 ex_stat LOCAL localhost   RW     0       INF      0       1
P proc2 ex_cmd  LOCAL localhost   RW     0       INF      0       2
P proc2 ex_stat LOCAL localhost   RW     0       INF      0       2
```

Chapter 8

Other Classes and Functions

CMS/NML classes of the RCS library were explained earlier. In this chapter we describe some other lower-level classes and functions which may be useful in developing RCS applications. These utilities include the RCS_TIMER class, some other time functions, the RCS_SEMAPHORE class, the RCS_LINKED_LIST class, the RCS print functions, and the Windows functions of the RCS library. To use these utilities it is enough to include rcs.hh and to link with the RCS library, just as you would do to use CMS/NML classes.

8.1 RCS_TIMER Class

In discrete-time control applications you need a timer for establishing the sampling frequency. Even in advanced intelligent and autonomous structures such as RMA (discussed in Chapter 3) there is still a need for time references. Recall that from the timing diagram of RMA (Figure 3.3), every module can have its own timing. For example, consider control of a chemical process. The time constants (i.e., the time required for the process to come to a steady state after a step input) for concentration, pH factors, and so on, are very small, in general, hence they require faster sample rates. On the other hand, the time constants for temperature or level control of the substances are large; hence they can tolerate slower sample rates. This shows the need for different cycle times for different parts of a single plant. In general, however, the timing requirements of the modules at a particular level in the RMA structure are similar and they are generally about an order of magnitude faster than the timing requirements of the modules one layer above in the architecture.

In conventional discrete-time control applications, cycle times are set, in general, using interrupt capabilities of the computers, using timer or counter capabilities of data acquisition boards, and so on. Therefore, in complex control systems with sophisticated timing requirements these tools may not be enough to satisfy all the timing needs. For this reason, we need more general means of establishing cycle rates. The RCS library provides the `RCS_TIMER` class, which was developed for this purpose. It can be used by the programmers to synchronize to the system clock or to some other event(s), for example, terminal count for a timer on the data acquisition board. In order to synchronize a cyclic process to the system clock you need to initialize the `RCS_TIMER` object with the cycle period and call `RCS_TIMER::wait()` at the end of each cycle. The cycle period will be rounded up to the resolution of the system clock, or the most precise time measuring or sleeping function available for the given platform. `RCS_TIMER::wait()` will wait the remainder of the cycle period since the last call. Example 8.1 shows how you can use the system clock as a synchronization event.

Example 8.1: Synchronization to the system clock

```
#include "rcs.hh"

main()
{
    RCS_TIMER timer(0.02);/* Initialize timer for 20 millisecond cycles. */

    while(1)
        {
            /* Do some processing. */
            timer.wait();        /* Wait for the end of cycle. */
        }
}
```

In order to synchronize to some other event(s) you need to create a function that takes `(void *)` as an argument and returns an `int`. After that, you need to initialize the `RCS_TIMER` object with a cycle period used only for diagnostics, the address of the function that was already created, and a parameter for the function. Having done that, the `RCS_TIMER::wait()` is used as in Example 8.1.

The user's function should return 0 when the event of synchronization occurs, or -1 if an error occurs. The argument passed to the user's function will be whatever was passed as the third parameter to the constructor of the `RCS_TIMER`, or `NULL` if no third argument is given. This argument could be used by the synchronizing function to know which timer is calling it if the synchronization function is called by more than one timer. Nothing will force the function to return within the cycle period, but there are ways to check if the function took longer than the cycle period after it returns.

Example 8.2 shows how the timer class can be synchronized to an event(s) different than the system clock. It implements a timer that waits until a key is pressed by the user.

Example 8.2: Synchronization to a character input from the keyboard

```
#include "rcs.hh"
#include <stdio.h>

int my_sync_func(void *arg)
{
    getchar();
    return(0);
}

main()
{
    RCS_TIMER timer(0.02, my_sync_func, NULL);

    while(1)
        {
            /* Do some processing. */

            timer.wait();
        }
}
```

The format of the constructor for initializing a new RCS_TIMER class object is

RCS_TIMER::RCS_TIMER(double *_interval*,
    RCS_TIMERFUNC *_function* = NULL, void **_arg* = NULL)

where *_interval* is the cycle period, *_function* is an optional function for synchronizing to an event other than the system clock, and *_arg* is a parameter that will be passed to that synchronizing function. The other member functions of the RCS_TIMER class are as follows:

- `int RCS_TIMER::wait();`
  Wait until the end of the interval or until a user function returns. Returns 0 for success, the number of cycles missed if it missed some cycles, or -1 if some other error occurred.

- `double RCS_TIMER::load();`
  Returns the percentage of loading by the cyclic process. If the process spends all of its time waiting for the synchronizing event, then it returns 0.0. If it spends all of its time doing something else before calling the `wait` function in a given cycle, then it returns 1.0. The load percentage is the average load over all of the previous cycles.

- `void RCS_TIMER::sync();`
  Restarts the wait interval.

The RCS_TIMER class of the RCS library makes the job of the programmer much easier because you do not have to worry about developing a code for setting up a sampling time. In the programs not using the timer you have to deal with the counters and the interrupt flags in the data acquisition card for setting up a sampling frequency, whereas with `RCS_TIMER`, setting cycles is much easier.

The `RCS_TIMER` class was already used in some of the examples in Chapters 5 and 6 (see Example 5.10 and Example 6.7). We recommend that the reader review these examples to gain better insights into how it is used within an RCS application program.

The feature of the timer to be synchronized to the end of an event, which is different from the system clock, are not used in the examples in this book. It was presented because it can be necessary and useful in other more complicated applications (e.g., ones that could be represented with a discrete event system model with asynchronous events).

8.2 Other Time Functions

In the RCS library, two more time functions are available, which can be useful in some applications. They can be used to put a process to sleep or to check elapsed time. These time functions are the following:

- `void esleep(double` *_secs*`);`
  Puts the calling process to sleep for *_secs* seconds. The time is rounded up to the resolution of the system clock, or the most precise sleep or delay function available for the given platform.

- `double etime();`
  Returns the number of seconds from some event. The time is rounded up to the resolution of the system clock, or the most precise time-measuring function available for the given platform. Generally, for the value returned to mean anything useful, you need to be able to compare it with a value stored from a previous call to `etime()`.

The programmer can find a variety of uses for these functions. For example, if we know that a particular process has to wait a certain amount of time before beginning a job, we can put this process to sleep using the function `esleep` so that the processor time is utilized more efficiently. On the other hand, you can use the function `etime`, for instance, to measure how long it takes the system to do a particular job. Such information may be needed in planning or for other purposes.

8.3 RCS_SEMAPHORE Class

Mutual exclusion is of paramount importance in systems using shared resources. (The reader who is not familiar with the concepts of multitasking and mutual exclusion should consult Appendix C or one of many books on operating systems, such as [15, 37].) For example, consider the shared memory buffers that the RCS library uses for communications. Assume that there is no mutual exclusion for the processes using the same buffer. If a reader process reads the contents of the buffer while a writer process has written only a part of it, the reader will get inaccurate information. This may lead to an improper control decision and therefore can be catastrophic. As another example, consider a manufacturing system where two robots grab the same tool because of lack of mutual exclusion. This may lead to a breakdown of the system.

The RCS library provides some tools that can be used for mutual exclusion of resources shared by multiple processes. In fact, a class called `RCS_SEMAPHORE` was developed to be used for mutual exclusion purposes. This class is supported in all the platforms except DOS and Windows (see Appendix E for a table of platforms tested). Mutual exclusion of the NML shared memory buffers is done by the RCS library (see Chapter 7 for details); however, the RCS application programmer may need to use the `RCS_SEMAPHORE` for other shared resources.

To use this utility you should create a semaphore or attach to an existing semaphore by initializing the `RCS_SEMAPHORE` object. The semaphore should have a unique identification which should be used by all the processes using the semaphore, and one of these processes should be responsible for creating this semaphore.

The two functions for taking and releasing the semaphore for mutual exclusion are `RCS_SEMAPHORE::wait()` and `RCS_SEMAPHORE::post()`, which should surround the "critical region." Example 8.3 shows how a semaphore object is created and how mutual exclusion is implemented.

Example 8.3: Mutual exclusion

```
/* utilex3.hh */
#ifndef UTILEX3_HH
#define UTILEX3_HH

#ifdef __cplusplus
extern "C" {
#endif
#include <sys/stat.h>    /* defines the S_IXXXX permission flags */
#ifdef __cplusplus
} /* END of extern "C" */
#endif

/* ID that processes connecting to this semaphore must agree on. */
#define MY_SEM_ID 101

/* Permissions Mode for rw_rw_r__ */
#define MY_SEM_MODE (S_IRUSR | S_IWUSR | S_IRGRP | S_IWGRP | S_IROTH)

#endif  /* end of UTILEX3_HH */

/* utilex3a.cc */
#include "utilex3.hh"
#include "rcs.hh"

/* Process A */
main()
```

```
{
    RCS_SEMAPHORE my_sem(MY_SEM_ID, /* Both processes must agree on id. */
                         RCS_SEMAPHORE_CREATE,    /* Create the semaphore. */
                         0.100,    /* timeout = 100 milliseconds */
                         MY_SEM_MODE,   /* Set permissions for semaphore */
                         1);          /* Initial State */

    // Some processing .....

    if(-1 != my_sem.wait())
        {
            /* Access shared resource. */
            my_sem.post();
        }

    // Some processing .....
}

/* utilex3b.cc */
#include "utilex3.hh"
#include "rcs.hh"

/* Process B */
main()
{
    RCS_SEMAPHORE my_sem(MY_SEM_ID, /* Both processes must agree on id. */
                         RCS_SEMAPHORE_NOCREATE, /* Don't Create semaphore.*/
                         0.100); /* timeout = 100 milliseconds */

    // Some processing .....

    if(-1 != my_sem.wait())
        {
            /* Access shared resource. */
            my_sem.post();
        }

    // Some processing .....
}
```

The format of the constructor for the RCS_SEMAPHORE class is

RCS_SEMAPHORE::RCS_SEMAPHORE(unsigned long int *_id*, int *_oflag*,
double *_timeout*, int *_mode* = DEFAULT_SEM_MODE, int *_state* = 0)

It initializes an RCS_SEMAPHORE object. The parameter *_id* should be a unique identifier for the semaphore which will be used by the processes accessing it. If RCS_SEMAPHORE_CREATE is passed to *_oflag*, then a semaphore is created. If RCS_SEMAPHORE_NOCREATE is passed *_oflag*, then the process will try to attach to a semaphore that must already have been created with the same *_id*. If *_timeout* is positive, then calls to RCS_SEMAPHORE::wait() will return -1 after *_timeout* seconds. If *_timeout* is negative, then RCS_SEMAPHORE::wait() will wait indefinitely for the semaphore to be available. If *_timeout* is zero, then RCS_SEMAPHORE::wait() will return immediately with 0 if the semaphore was available or -1 if it was not. The *_mode* determines which users will have permission to use the semaphore. The default value of *_mode*, `DEFAULT_SEM_MODE`, allows read and write access to every process which has access to the semaphore. The *_mode* will be ignored if the process is not creating the semaphore. The *_state* should be 1 to make the semaphore immediately available. The *_state* will be ignored if the process is not creating the semaphore.

The other function members of the `RCS_SEMAPHORE` class are as follows:

- `int RCS_SEMAPHORE::wait();`
  Wait for the semaphore to be available and then take it. See the constructor's parameters for several options affecting its behavior. Returns 0 for success, or -1 for failure.

- `int RCS_SEMAPHORE::trywait();`
  If the semaphore is available, take it. Returns 0 for success, or -1 for failure.

- `int RCS_SEMAPHORE::post();`
  Release the semaphore. Returns 0 for success, or -1 for failure.

- `int RCS_SEMAPHORE::getvalue();`
  Test to see if the semaphore is available, but don't take it even if it is. Returns a positive integer if the semaphore is available, or 0 if it is not.

As mentioned earlier, if you don't want to deal with mutual exclusion of the shared memory, you can leave this job to the RCS library. (See Chapter 7 for details on mutual exclusion of shared memory.) This can be done by specifying the option `mutex=os_sem` in the NML configuration file. This option will also be chosen by default if the programmer does not specify any type of mutual exclusion. In that case the RCS library will automatically set up an operating system semaphore for mutual exclusion for the shared memory. We recommend that the programmer try to leave the mutual exclusion to the RCS library if some other kind of mutual exclusion is not required. Still, it is good to know about the `RCS_SEMAPHORE` because in some applications, mutual exclusion constrains other than constraints on the shared memory, may arise. For example, there can be other shared resources, such as tools which are used by different subsystems of the RCS application. In that case, the programmer may use the `RCS_SEMAPHORE` class to make sure that the shared resource is accessed properly.

8.4 RCS_LINKED_LIST Class

Linked lists are very useful tools for dynamic memory allocation. That is, in linked lists, in contrast to arrays, the programmer does not allocate all the memory in the beginning of the program. However, new nodes of the list are created whenever necessary and removed when not needed.

Linked lists can be used for a variety of applications. For example, they can be useful in the planning processes of the modules in the RMA structure. There can be different possible future actions which can be generated and stored as tentative plans in a linked list until the optimal plan is found. Then all other plans may be removed from the memory. Since each plan may involve a different number of events, the memory requirements for a plan may not be known a priori; there may be a need for dynamical allocation of the memory. Other uses of linked lists in RMA may involve storing the events and the entities in the *knowledge database* of the system, where new events encountered can be added to the database, and those determined not to exist anymore may be removed.

The RCS library has a class, called RCS_LINKED_LIST, for providing a linked list utility. It includes an internal pointer that keeps track of the current node. In order to use the linked list, the programmer first needs to initialize it. After that, you can store some objects in the list and perform operations needed on the objects. Example 8.4 illustrates how you can use this utility.

Example 8.4: Linked list

```
#include "rcs.hh"

class MY_STRUCT {
public:
    int count;
};

RCS_LINKED_LIST *my_list;
int compute_total_count(RCS_LINKED_LIST *my_list);

main()
{
    int totalCount;
    MY_STRUCT S1, S2, S3;
    my_list = new RCS_LINKED_LIST();/* Initialize and create linked list. */

    /* Store S1 on the end of list. */
    my_list->store_at_tail(&S1, sizeof(MY_STRUCT), 0);

    /* Store S2 on the end of list. */
    my_list->store_at_tail(&S2, sizeof(MY_STRUCT), 0);

    /* Store S3 on the end of list. */
    my_list->store_at_tail(&S3, sizeof(MY_STRUCT), 0);

    totalCount = compute_total_count(my_list);
}

int compute_total_count(RCS_LINKED_LIST *my_list)
{
    MY_STRUCT *ptr_to_struct;
    int total_count = 0;

    /* Get first object and initialize internal pointer to start of list. */
    ptr_to_struct = (MY_STRUCT *) my_list->get_head();

    while(NULL != ptr_to_struct)
        {
```

```
            total_count += ptr_to_struct->count;
            ptr_to_struct = (MY_STRUCT *) my_list->get_next();
        }
    return(total_count);
}
```

The member functions of the RCS_LINKED_LIST class are:

- `RCS_LINKED_LIST::RCS_LINKED_LIST();`
  This is the constructor of the class. It initializes the linked list.

- `int RCS_LINKED_LIST::store_at_head(void *`*_data*`, size_t `*_size*`, int `*_copy*`);`
  This function creates a new node and places it at the beginning of the list. If *_copy* is nonzero, then this function will allocate `_size` bytes and copy *_size* bytes from the address starting at *_data* there, and then the `get` functions will return a pointer to the copy of the object. If *_copy* is zero, then the *_data* pointer will be stored and the `get` functions will return a pointer to the original object.

  If successful, it returns a positive integer *id* that can be used to select this node later, or -1 if an error occurred.

- `int RCS_LINKED_LIST::store_at_tail(void *`*_data*`, size_t `*_size*`, int `*_copy*`);`
  This function creates a new node and places it at the end of the list. If *_copy* is nonzero, then this function will allocate *_size* bytes and copy *_size* bytes from the address starting at *_data* there, and the `get` functions will return a pointer to the copy of the object. If *_copy* is zero, then the *_data* pointer will be stored and the `get` functions will return a pointer to the original object.

  If successful, it returns a positive integer *id* that can be used to select this node later, or -1 if an error occurred.

- `int RCS_LINKED_LIST::store_before_current_node(void *`*_data*`, size_t `*_size*`, int `*_copy*`);`
  This function creates a new node and places it before the current node. If *_copy* is nonzero, then this function will allocate *_size* bytes and copy *_size* bytes from the address starting at *_data* there, and the `get` functions will return a pointer to the copy of the object. If *_copy* is zero, then the *_data* pointer will be stored and the `get` functions will return a pointer to the original object.

  If successful, it returns a positive integer *id* that can be used to select this node later, or -1 if an error occurred.

- `int RCS_LINKED_LIST::store_after_current_node(void *`*_data*`, size_t `*_size*`, int `*_copy*`);`
  This function creates a new node and places it after the current node. If

_copy is nonzero, then this function will allocate *_size* bytes and copy *_size* bytes from the address starting at *_data* there, and the `get` functions will return a pointer to the copy of the object. If *_copy* is zero, then the *_data* pointer will be stored and the `get` functions will return a pointer to the original object.

If successful, it returns a positive integer *id* that can be used to select this node later, or -1 if an error occurred.

- `void *RCS_LINKED_LIST::get_head();`
  Get the address of the first object on the list and set the current node to the beginning of the list. If the list is empty, `get_head` returns `NULL`. Depending on how the object was stored, the address this function returns may be the address of the original object or of a copy.

- `void *RCS_LINKED_LIST::get_tail();`
  Get the address of the object at the end of the list and set the current node to the end of the list. If the list is empty, `get_tail` returns `NULL`. Depending on how the object was stored, the address this function returns may be the address of the original object or of a copy.

- `void *RCS_LINKED_LIST::get_next();`
  Get the address of the next object on the list and move the current node one step closer to the tail. If the list is empty, `get_next` returns `NULL`. Depending on how the object was stored, the address this function returns may be the address of the original object or of a copy.

- `void *RCS_LINKED_LIST::get_last();`
  Get the address of the previous object on the list and move the current node one step closer to the head. If the list is empty, `get_last` returns `NULL`. Depending on how the object was stored, the address this function returns may be the address of the original object or of a copy.

- `void RCS_LINKED_LIST::delete_current_node();`
  Remove the current node from the list and free any memory associated with it. Some extra pointers keep track of the node that was before and after the deleted node so that the next call to `get_next` or `get_last` will return the same object as if the current node was not deleted.

- `void RCS_LINKED_LIST::delete_node(int` *_id*`);`
  Delete the node with the associated *_id*.

- `void RCS_LINKED_LIST::set_list_sizing_mode(int` *_maximum_size*`, LIST_SIZING_MODE` *_new_mode*`);`
  Sets a sizing mode and the maximum number of nodes allowed on the list. The sizing mode determines what happens when there is an attempt to add another node to the list after it has reached the *_maximum_size*. The following are the possible values for *_new_mode*:

  - `DELETE_FROM_TAIL`: Remove one node from the tail of the list to make room for the new node.
  - `DELETE_FROM_HEAD`: Remove one node from the head of the list to make room for the new node.
  - `STOP_AT_MAX`: Return -1 if an attempt is made to add a new node when the list is full.
  - `NO_MAXIMUM_SIZE`: Allow the list to grow until all available memory is used up.

- `RCS_LINKED_LIST::~RCS_LINKED_LIST();`
  This function is the destructor of the class. It removes every node from the list and frees all the memory associated with the nodes or the list itself.

8.5 RCS Print Functions

Although ANSI C provides input/output facilities such as **printf**, these functions may not be appropriate under certain circumstances. For example, under Windows unless you compile with the EasyWin (Borland C++) or QuickWin (Visual C++) option, there will be no window created to catch **printf** messages, and there are some limitations if you choose the EasyWin or QuickWin options. The RCS print functions work much like the ANSI C facilities but can more easily redirect their output. Controlling where the output of the RCS print functions goes also controls the error messages of CMS/NML. They are supported by all the platforms shown in Appendix E. Next, we list these functions with a short description of their use.

- `int rcs_print(char *`*_fmt*`, ...);`
  Prints a message using the *_fmt* format string and optional additional arguments using the `printf` conventions.

- `int rcs_print_error(char *`*_fmt*`, ...);`
  Prints a message using the *_fmt* format string and optional additional arguments using the `printf` conventions if the `PRINT_RCS_ERRORS` flag is set. (See `set_rcs_print_flag()` function below.)

- `int rcs_print_debug(long `*_flag_to_check*`, char *`*_fmt*`, ...);`
  Prints a message using the *_fmt* format string and optional additional arguments using the `printf` conventions if the corresponding debug flag to *_flag_to_check* is set. (See `set_rcs_print_flag()` function below.)

- `int rcs_vprint(char *`*_fmt*`, va_list `*_va_args*`);`
  Prints a message using the *_fmt* format string and the *_va_args* using the `vprintf` conventions.

- `int rcs_puts(char *`*_str*`);`
  Prints the string *_str* and adds a new line character at the end following the `puts` convention.

- `char *strip_control_characters(char *`*_dest*`, char *`*_src*`);`
  Removes new lines, carriage returns, and tabs from the *_src* string and stores the result in the *_dest* string if the *_dest* pointer does not equal `NULL`. If the *_dest* pointer equals `NULL`, the new string is stored in an internal array. Returns the *_dest* pointer or the address of the internal array where the new string was stored.

- `void set_rcs_print_destination(RCS_PRINT_DESTINATION_TYPE` *_type*`);`
  Changes where the output of the `rcs_print` functions is directed. The following choices are available:

  - `RCS_PRINT_TO_STDOUT` – Print to stdout.
  - `RCS_PRINT_TO_LOGGER` – Currently prints to stdout, except under VxWorks, where it uses the `logMsg` function, which is nonblocking.
  - `RCS_PRINT_TO_STDERR` – Print to stderr.
  - `RCS_PRINT_TO_NULL` – Make all `rcs_print` functions return without doing anything.
  - `RCS_PRINT_TO_LIST` – Store all `rcs_print` messages in a linked list, so that later they can be displayed in a separate window, or used in some other way. The current list sizing mode defaults to a maximum size of 256 with excess nodes being deleted from the head.

- `void set_rcs_print_flag(long` *flag_to_set*`);`
  An internal 32-bit integer contains a set of flags that are checked whenever an `rcs_print_debug` or `rcs_print_error` occurs, to determine whether or not the message should be printed. Programmers can define their own flags in the most significant byte or turn on or off several CMS/NML debug messages.

  The current messages that can be turned on or off are:

  - `PRINT_RCS_ERRORS` /* ON by default.*/
  - `PRINT_NODE_CONSTRUCTORS`
  - `PRINT_NODE_DESTRUCTORS`
  - `PRINT_CMS_CONSTRUCTORS`
  - `PRINT_CMS_DESTRUCTORS`
  - `PRINT_NML_CONSTRUCTORS`
  - `PRINT_NML_DESTRUCTORS`
  - `PRINT_COMMANDS_RECIEVED`

  - PRINT_COMMANDS_SENT
  - PRINT_STATUS_RECIEVED
  - PRINT_STATUS_SENT
  - PRINT_NODE_CYCLES
  - PRINT_NODE_MISSED_CYCLES
  - PRINT_NODE_CYCLE_TIMES
  - PRINT_NODE_PROCESS_TIMES
  - PRINT_NEW_WM
  - PRINT_NODE_ABORT

- `void clear_rcs_print_flag(long` *flag_to_clear*`);`
  Clears a flag set with `set_rcs_print_flag`.

- `RCS_LINKED_LIST *get_rcs_print_list();`
  Returns the address of the linked list where messages may have been stored.

- `void clean_print_list();`
  Deletes the linked list where messages may have been stored.

- `void set_rcs_print_list_sizing_mode(int` *max_size*`, LIST_SIZING_MODE` *mode*`);`
  The print list will have a default maximum size of 256 and it will delete nodes from the head of the list to make room for new nodes. To change these settings it is preferable to use this function rather than getting the print list and using `RCS_LINKED_LIST::set_list_sizing_mode()`, since it works regardless of whether the print list has been initialized yet.

Some of the RCS print functions such as `rcs_print` were used instead of standard C/C++ print functions for printing messages to the screen in the development of the code for the tank example discussed in Chapter 2.

8.6 Windows Functions

The functions listed in this section are supported by MS-Windows 3.11, Windows 95, and Windows NT with Borland C++, or Microsoft Visual C++. They are used to print the messages to a given window. After the messages from `rcs_print` have been stored in the linked list, Windows programs can go through the list themselves or create a special window to display these messages automatically.

To create a window for the `rcs_print` messages you need to create a print window. This can be done using `create_rcs_print_window()` function. Its format is

HWND create_rcs_print_window(HANDLE *hInstance*, int *nCmdShow*, HWND *hwndParent*)

This function creates a window to display `rcs_print` messages. It is updated automatically when the `rcs_print` function is called or when one of its window's controls are used. The *hInstance* parameter should equal the first parameter passed to WinMain. The `nCmdShow` will be passed to `ShowWindow`. See the Windows API for ShowWindows options. The two most common values of *nCmdShow* are `SW_MINIMIZE` to start the window as an icon or `SW_SHOWNORMAL` to open the window immediately. You can create the window as a child of *hwndParent* or set *hwndParent* to `NULL` to make the window independent.

When finished with the print window, you can remove this window by using

```
void remove_rcs_print_window()
```

which sends the `WM_DESTROY` message to the rcs print window.

We finish the section with Example 8.5, which shows the use of print and window functions.

Example 8.5: Printing under Windows

```
/* roots.hh */
#ifndef ROOTS_HH
#define ROOTS_HH

/* Function for computing the roots of a quadratic. */
void compute_roots(double A, double B, double C, double &root1, double &root2);

#endif /* end of ROOTS_HH */

/************************************************************/

#include "roots.hh"
#include "rcs.hh"
#include <math.h>

/* Function for computing the roots of a quadratic. */
void compute_roots( double A, double B, double C, double &root1, double &root2)
{
    if(B*B-4*A*C < 0)
        {
            rcs_print("compute_roots: Can't compute square root of %lf\n",
                      B*B-4*A*C);
            return;
        }
    root1 = (-B+sqrt(B*B-4*A*C))/(2*A);
    root2 = (-B-sqrt(B*B-4*A*C))/(2*A);
}

/************************************************************/

#include "rcs.hh"
#include "roots.hh"
```

```
/* A UNIX or DOS Application which uses compute_roots. */
main()
{
    double r1, r2;
    /* Send all the rcs_print messages to the standard output. */
    set_rcs_print_destination(RCS_PRINT_TO_STDOUT);

    /* Try to compute the roots of a quadratic that will cause an error.*/
    compute_roots(1, 1, 1, r1, r2);
}

/***********************************************************/

#include "rcs.hh"
#include "roots.hh"
#include <windows.h>

/* A Windows Application which uses compute_roots. */
int PASCAL WinMain( HANDLE hInstance,
                    HANDLE hPrevInstance,
                    LPSTR lpszCmdParam,
                    int nCmdShow)
{
    double r1, r2;
    MSG msg;

    /* Store all the rcs_print messages in a linked list
       that a window can display later. */
    set_rcs_print_destination(RCS_PRINT_TO_LIST);

    /* Create a window to show rcs_print messages,
       and display it as an icon to start.
       It has no parent window. */
    create_rcs_print_window(hInstance, SW_MINIMIZE, NULL);

    /* Compute the roots of a quadratic that will produce an error. */
    compute_roots(1, 1, 1, r1, r2);

    /* Wait for someone to kill this process. */
    while(GetMessage(&msg, NULL, 0, 0))
        {
            TranslateMessage(&msg);
            DispatchMessage(&msg);
        }
    return(msg.wParam);
}
```

The programs for the controllers of the application examples discussed later are run under either DOS or Linux operating systems. Therefore, there is no need for the use of the Windows print functions. They can be useful in some applications which run under operating systems such as Windows NT.

Chapter 9

RCS Diagnostics Tool

In the preceding chapters we discussed the basic elements of the RCS library. In this chapter, and in the next chapter, we focus on tools developed by NIST which are not directly in the library; however, they are of paramount importance for RCS application development. These are the RCS diagnostics tool which we discuss in this chapter and the RCS design tool, which is a subject in the next chapter.

The RCS diagnostics tool is a graphical interface developed in Java and can be viewed using any Web browser which supports Java, such as *Netscape* or *Microsoft Internet Explorer*, or it can be run as a stand-alone application provided that the JDK for the given platform is installed in the system. The purpose of the diagnostics tool is to provide a high-level interface to the control routines. It is through the diagnostics interface that the human operator can read state values of the plant that reflect performance and send commands to controllers to perform specific operations. It allows programmers to see the current command and status of every module in the controller and to send commands to any module. Commands which each module will accept, and the status and command message parameters of each module, are determined by using some of the C++ header files the application is built with.

The RCS diagnostics tool reads the information about the structure of an RCS controller from an architecture file, which is a text file that provides the controller structure, the command, status, and auxiliary buffers, TCP communication port numbers for accessing these buffers, and so on. The RCS design tool also uses the architecture files for obtaining information about an RCS controller. Architecture files can be generated automatically by the RCS design tool and are discussed in more detail in Chapter 11.

As we mentioned above, the diagnostics tool uses the same C++ header files that the application was built with. However, there are some issues which the programmer should be aware of while developing the code for header files so that they be compatible with the diagnostics. In the next section we discuss the rules for the header files.

9.1 Special Rules for the Header Files

The diagnostic tool uses the C++ header files of an application to read the declaration of command and status messages. This feature of the tool makes it easy to add new variables to the command and status messages and to add new command and status messages. That is, the programmer need not worry about making the same changes in both controller code and diagnostics code because both of these are using the same files. However, you should note that the tool is not a C++ compiler, so you should be careful about the following issues.

1. `#include` directives are ignored. Therefore, you should use the attribute `predefined_types_file` in the architecture file if you need some header files of predefined types to be included.

2. `#if` directives are considered false unless `JAVA_DIAG_APPLET_FORCE_TRUE` is defined.

3. Only one variable should be defined on a line.

4. The things that should not be parsed by the diagnostic tool should be surrounded with `#ifndef JAVA_DIAG_APPLET` and `#endif`. This includes the following:

    - Multiple line functions and macros.
    - Variables included in the message but not updated in the NML update function.
    - Large classes which will never be sent as NML messages and will just slow down the tool and use more memory to include.

5. The order of variables in their declaration should match the order in the update functions.

6. Each NML message should have a `#define` statement in the same header file giving the *message type id* formed by adding `_TYPE` to the class name. For example, if a class called `DRIVE_CHANGE_LANE` is defined, there should be the line `#define DRIVE_CHANGE_LANE_TYPE 4002` also. The number here is a unique type identifier.

7. Enumerated data types cause a problem for the NML update function, but the diagnostics tool can use them. Therefore, in the definition of enumerated data types you may use `#if`'s so that these definitions are read as enumerated by the diagnostic tool, but as regular data by the NML update functions. The code in Example 9.1 provides an example for this. In this example the code will make the communications work as before but the tool will display `WMSA_AUTO`, `WMSA_MDI`, and `WMSA_MANUAL` instead of 0, 1, and 2, respectively.

Example 9.1: Using enumerated data in the RCS diagnostics tool

```
enum WMSA_MODE
{
    WMSA_AUTO,
    WMSA_MDI,
    WMSA_MANUAL
};

class NML_WMSA_WM: public NMLmsg
{
    . . .
#ifdef JAVA_DIAG_APPLET
    WMSA_MODE mode;
#else
    int mode;
#endif
    . . .
};
```

We recommend that the programmer keep in mind these rules for the header files while developing code. In the next section we explain how you can use the diagnostics tool and provide some of the possible views of it for better illustration.

9.2 Using the Diagnostics Tool

As mentioned above, to use the RCS diagnostics tool you need either a Web browser that supports Java or the Java Developer's Kit installed on the system. If you are using a browser which has a just-in-time (JIT) compiler, the applet will run much faster. To start the applet, you need to open the Web browser and to load the HTML file for it. (We describe the preparation of an HTML file in the next section.) After the applet is started, depending on some parameters set in the HTML file, it will either load an architecture file and connect to a running controller or wait for the user to enter the name or URL of an architecture file in the box "Hierarchy URL" on the top of the screen, and press the "LOAD" button to load it. Under this URL box there is a check box which, if checked, forces the tool to connect to the controller specified in the architecture file. If it is not connected, the text "NOT CONNECTED" appears next to it. Whenever checked, the text will eventually say "CONNECTED (n out of m)," where $n \leq m$. Above, n is the number of modules which connected successfully and m is the total number of modules. If there are some modules that did not connect, you can check the Java console or `log` file for some explanation of why a module did not connect.

Figure 9.1 shows a possible appearance of the diagnostics tool. On the left, there is a checkbox which allows users to choose to use color or not. The colored option may slow the system a little bit. The "FIND" button and the box next to it can be used to search and find the files that the user is looking for. For example, if the location of the configuration file is not known, you may search for it using this utility of the tool. There are other buttons, such as "Browse Local Files," "Stop all Plotting," and "Refresh" (see Figure 9.1). The "Browse Local Files" button is used to browse through the directories and files of the local computer when you want to load a new architecture file. "Stop all Plotting" is used to stop the plotting of all the variables that you have previously set for plotting. Later we will see how we can plot variables. You can use the "Refresh" button if you want the tool to reread the information from all the buffers. On the right of the screen, there is also an option to set up the refresh time for the screen. The response of the applet to pressing a button, for example, depends on this time. Setting a faster time can be better; however, it also requires more processor overhead. Under the checkbox for connecting there is a bar showing which file is currently loaded and the progress achieved.

Below these, on the left, there is a drop-down box with view options to choose from. See Figure 9.1, where currently this box shows `Login`. The appearance of the applet under this box depends on the option chosen. The available options are explained below, where we also provide some of the screen appearances for clarity.

- `Login`
  The appearance of the screen under this option is shown on Figure 9.1. This option is needed only if there is a need to prevent someone from

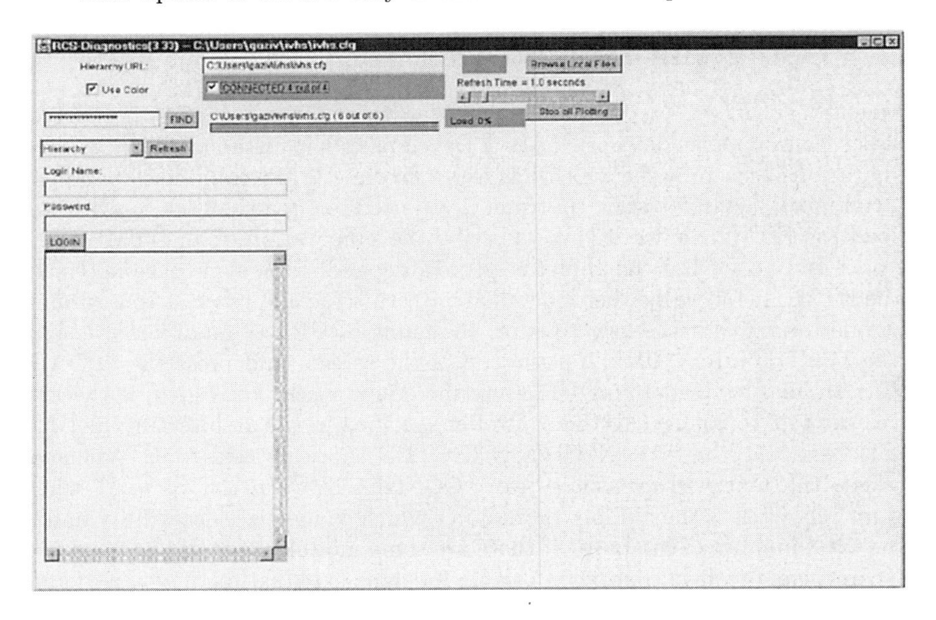

Figure 9.1: Login view of the applet.

having complete access to the NML buffers. To do this, you can add `passwd=file_name.pwd` to the end of each buffer line in the NML configuration file and after that create or modify the `.pwd` file using the program `nmlpwd`.

You will not be able to send commands to or see the status of buffers for which a password has been set unless you go to this option of the diagnostics tool and enter a valid `username` and `passwd`. It is also possible to allow some users only read access.

- `Details`
  If you choose the `Details` option, a table with the fields *Modules, Commands Available*, and *Cmd To Send* appears, as shown in Figure 9.2. The

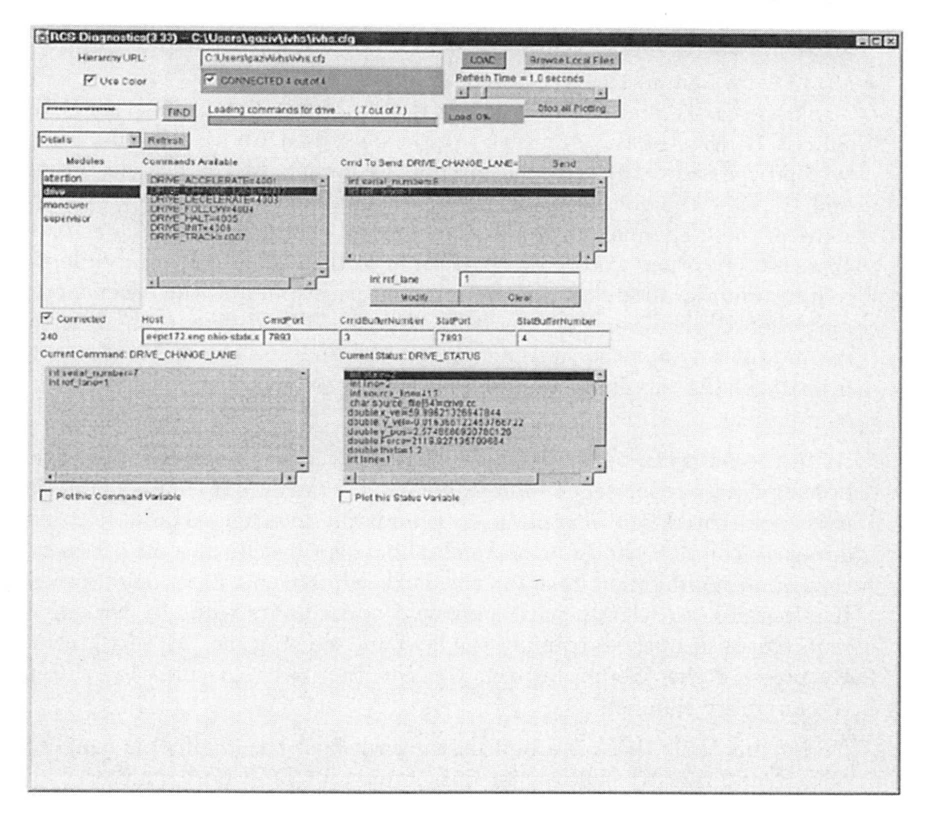

Figure 9.2: Details view of the applet.

column of modules shows all the modules in the controller that were loaded by the tool. If you choose a particular module by clicking on it, then on the column of available commands the commands which can be sent to this module appear. Selecting one of the available commands shows the parameters for that command under the command to send. You can edit commands by selecting them and then entering appropriate values in the text box for modification. Next, you can send the chosen command by pressing the "Send" button.

Below these, there are fields showing the host name, TCP port numbers for the command and status ports, and the buffer number for the command and status buffers. These are the values read from the architecture file. If there is a discrepancy between these values and the values in the NML configuration file, the programmer needs to change the numbers in the architecture file to match these in the NML configuration file.

The area under these fields is for viewing the fields of the current command and status messages. It is possible to plot one or more fields of the command and status messages by choosing a particular variable and then clicking the checkbox for plotting this control or status variable.

- `Auxiliary Channels`
  The appearance and function of this option of the diagnostics tool is very similar to that of the `Details` option described above. Its purpose is to send messages through the auxiliary channels in the system (if any) and to view the messages of the other auxiliary channels. (Note that some of these channels may act as command channels and some as status channels. In other words, through some of these channels a module may request another module to perform some operation, and the other module may report whenever the operation is done. The difference here is that the modules do not have a supervisor–subordinate relation. Moreover, the priority of the auxiliary request may be lower than that of a supervisor command.)

  As in the `Details` option, a table with fields `Aux. Channels`, `Available Message`, and `Message to Sent` appears. If a particular channel is chosen, it is highlighted, and on the next column the messages available to send appear. You can choose a particular message by clicking on it. In that case it is highlighted, and on the next window its data fields appear. It is possible to select a particular field and edit its value by writing the required value in the box under the field and clicking the "Modify" button. By pressing the "Send" button, you can then sent the message through the auxiliary channel.

  Under this table there is a box showing whether the channel is connected or not, and two other boxes for host name and TCP port number.

  Below these boxes there is a table for viewing the fields of the auxiliary messages. You can choose a given message by clicking on it. It is possible to plot any field of any message by choosing it and clicking the clickbox labeled "Plot this Variable."

- `Hierarchy`
  The appearance of the applet whenever this option is chosen is as shown in Figure 9.3. It shows the hierarchical construction of the controller. That is, all the modules are shown in a hierarchical order, and there are solid lines connecting the children to their parent modules and dashed lines

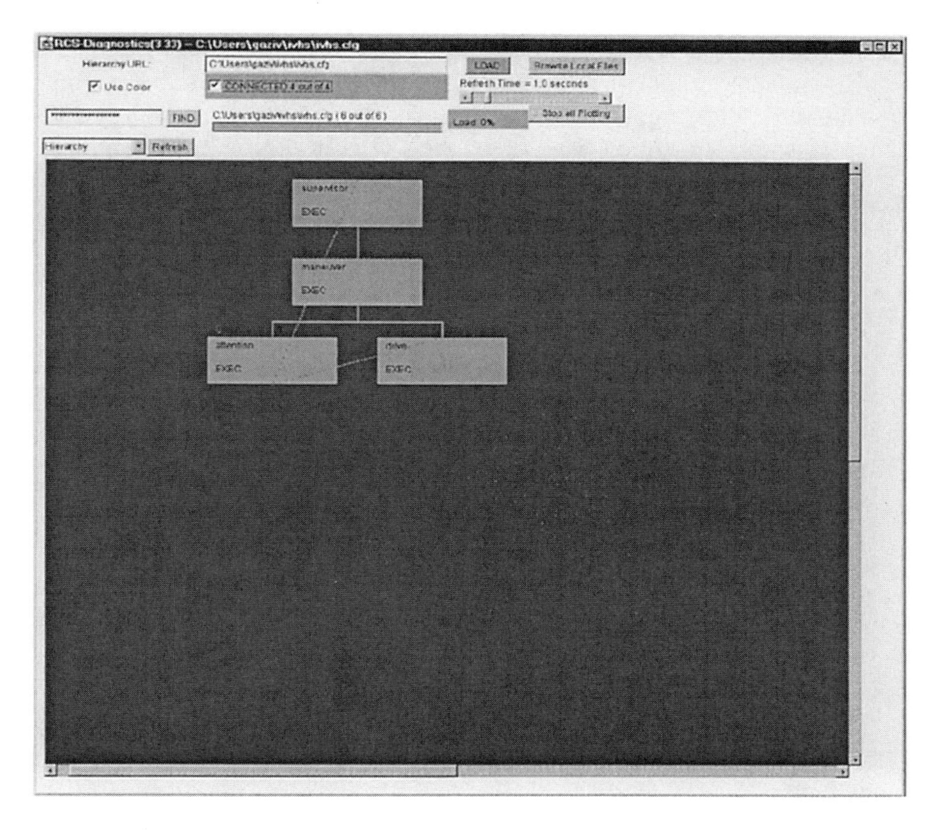

Figure 9.3: Hierarchy view of the applet.

showing the auxiliary channels. Each module box contains the name of the module, its current command, and its status. If the "Use Color" box in the top of the file is checked, then the modules are color coded. The meaning of the colors are as follows:

- `WHITE=DONE` – The module has completed all the commands it was given, or the status was unknown.
- `GREEN=EXECUTING` – The module is still working on the current command.
- `RED=ERROR` – The module is reporting some internal error.
- `GREY=NOT CONNECTED` – The tool is unable to communicate with this module.

In case there are modules which are not connected or modules reporting an error, the user may check the `error log` for messages in order to find the origins for them.

- `Graph`
  The `Graph` option is for plotting variables of the status and/or command messages. The appearance of the screen whenever this option is chosen is as shown in Figure 9.4. If, in the `Details` option, you have clicked

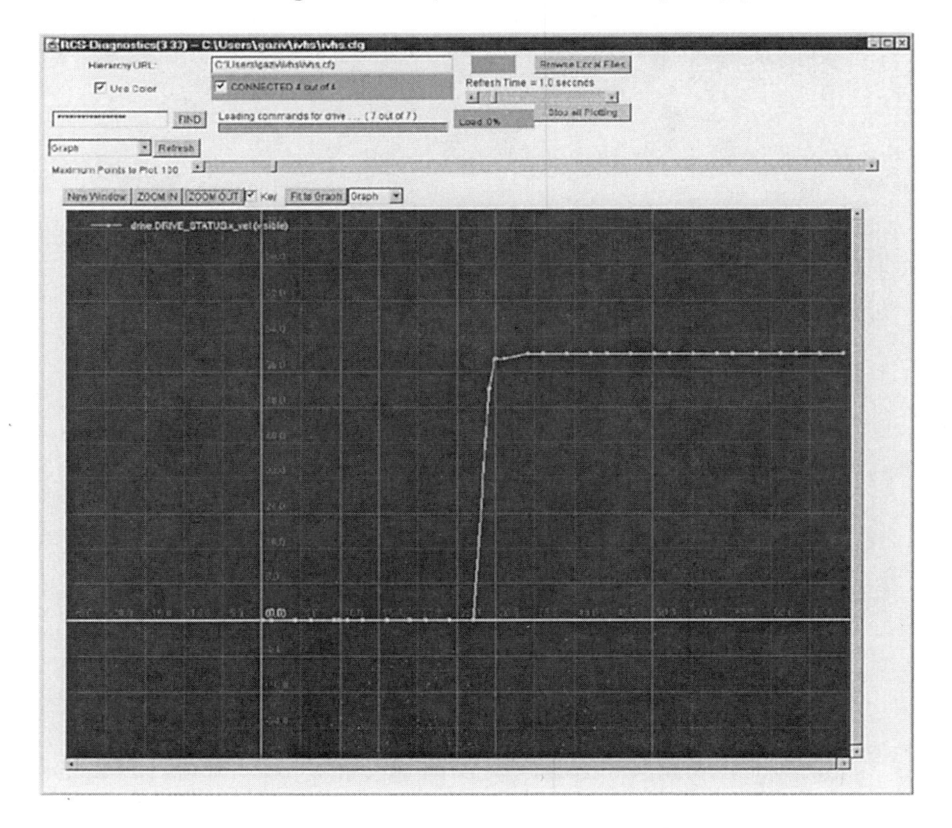

Figure 9.4: Graph view of the applet.

the checkbox for plotting a particular field of a command and/or status message, then these fields are plotted to the graph. This option has several suboptions which can be chosen from the drop-down box next to the text "Fit to Graph," which currently shows `Graph` in Figure 9.4. These are:

  - `Graph` – for plotting the variables.
  - `Text` – for observing the settings of the graph, such as ranges of the axes.
  - `Options` – for observing the options for the settings of the graph and changing them.
  - `Key` – for observing the key for the plots on the graph or which curve corresponds to which variable.

- `Error Log`
  If this option is chosen, a screen showing the messages written to the `error log` is displayed. The `error log` is a special NML buffer that any module can write to. The text of messages written there will be appended to the text area beneath the box. Also, a special `NML_DISPLAY` message can be sent to cause the browser showing the applet to jump automatically to a certain URL. This can be used for providing special instructions to an operator. Instead of telling the operator to load tool X with just text, it can jump the operator to a page with a picture of the tool and formatted instructions on how to load it. The appearance of the screen under this option is shown in Figure 9.5.

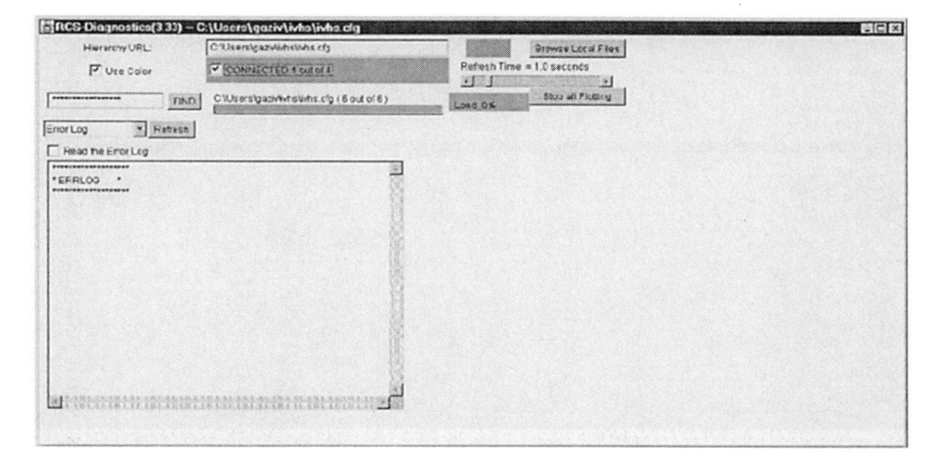

Figure 9.5: Error Log view of the applet.

- `State Table`
  State tables are user-defined methods of performing commands in the control modules (i.e., in classes derived from `NML_MODULE`). They are coded as `if-then-else` clauses which are called cyclically and implement a state machine. This is done using `STATE_MATCH` and `StateNext` functions of the `NML_MODULE`. The code should be developed so that every `if` statement corresponds to a particular state and only one `if` holds at a time. This option of the diagnostics tool shows the place within the state table which is currently executed, or in other words, the `if` statement which is currently true. The appearance of the screen under this option is shown in Figure 9.6.

- `Debug Flags`
  This option is for setting some debug flags. Setting a debug flag will cause the tool to print error messages to the `Error Log` buffer, which can be used by the programmer to find the source of an error or malfunction in case of any. The screen for this option is shown in Figure 9.7.

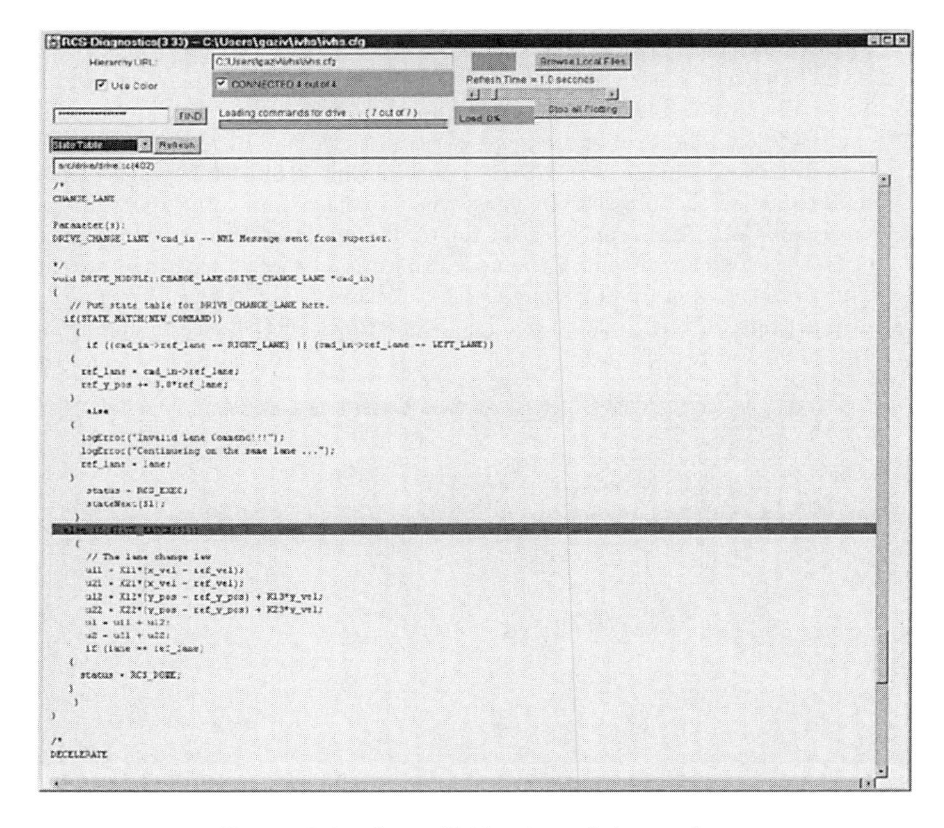

Figure 9.6: State Table view of the applet.

The meaning of the available debug flags is as follows:

– `Debug NML reads`
  Reports any errors occurring during NML read operations.

– `Debug NML writes`
  Reports any errors occurring during NML write operations.

– `Debug NML configs`
  Reports any NML configuration errors, such as two masters set for the same buffer.

– `Debug diagapplet`
  Reports any errors related to the RCS diagnostics tool itself.

– `Debug hierarchyPanel`
  Reports any errors or discrepancies in the controller hierarchy.

– `Debug plotter`
  Plotter is the RCS tool that plots the data on the `Graph` view of the diagnostics tool. This flag causes any errors related to this tool to be reported.

- `Debug codegen`
  The NML code generator is another RCS library tool that can be used to generate C++ code for the NML messages, and it will be discussed in Chapter 10. This flag causes any errors related to this tool to be reported.
- `Debug moduleinfo`
  Report any errors related to any of the RCS modules in the system.
- `Debug fileloader`
  Report any error occurring during opening or loading of files to the tool. Recall that the tool loads the code for the messages and the state tables so that the user can view the data fields of the messages and the state tables.
- `Debug DiagNMLMessageDictionary`
  Report any errors in the NML message dictionary, such as two messages with the same type number, or messages of unknown type.
- `Debug DiagNMLmsg`
  Report any errors in the NML messages.
- `Debug XDR (eXternal Data Representation) Converter`
  If you have established in the NML configuration file that the information in the buffers should be stored in the `XDR` format, then the messages will be converted to `XDR`. This option reports any errors occurring during this operation.

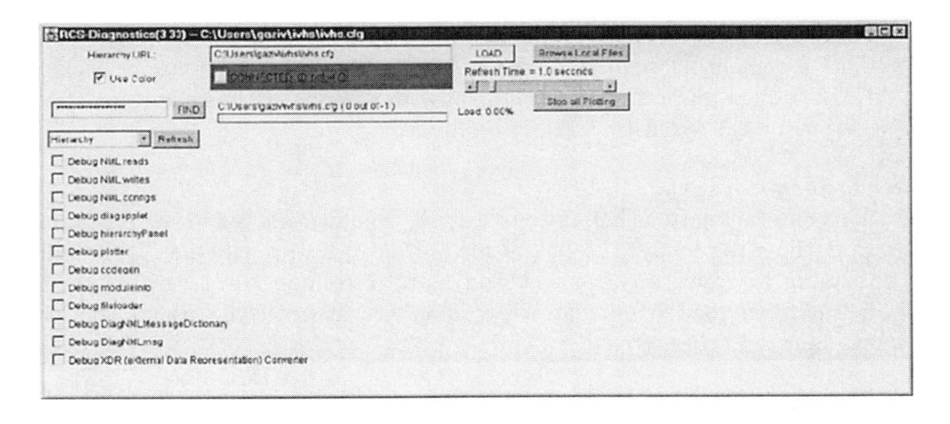

Figure 9.7: Debug Flags view of the applet.

9.3 Executing the Diagnostics Tool

There are three different ways to execute and view the RCS diagnostics tool. If you have the JDK installed in your system, you can execute it as a stand-alone application using command similar to the following:

```
java -classpath $CLASSPATH diagapplet.diagapplet
```

where the environmental variable should be properly set to the directory of the Java and RCS Java libraries. Normally, the Java libraries are located under `/usr/local/jdk/lib`, and usually the RCS Java libraries are installed under `/usr/local/rcslib/plat/java/lib` on UNIX or Linux systems. On Microsoft Windows systems, on the other hand, they could be located under, for example, `c:\Program Files\jdk\lib` and `c:\Rcslib\plat\java\lib`, respectively.

Two other options would be to view the tool as an applet using either

```
appletviewer diagapplet.html
```

or a Web browser to load the HTML file prepared for it. In both of the last two cases you have to prepare an HTML file for it.

There are several parameters that can be provided to the diagnostics tool on start. If you are executing the tool as a stand-alone application, you can provide these parameters on the command line, whereas if you are running the tool as an applet, you can provide them in the HTML file. Depending on these parameters, the applet will load the architecture file and try to connect to the controller or wait for you to enter the name of the architecture file to load. The following is a list of the available parameters.

- `hierarchyFile`
  The URL of a hierarchy configuration file to be loaded automatically.

- `refreshTime`
  This is a floating point number specifing the interval between checking NML buffers and refreshing the display in seconds. It has the same effect as moving the "Refresh Time" scrollbar.

- `connectOnStartup`
  This can be set to either `true` or `false`. The default is `false`. Assuming that the `hierarchyFile` is specified, this indicates whether the applet schould automatically connect and start refreshing the screen after the heirarchy is loaded or whether it should wait until the user clicks the "Connected" checkbox.

- `minimalMode`
  This can be set to either `true` or `false`. The default is `false`. Do you want to disable some features in order to save memory?

- `useColor`
  You can set this option to either `true` or `false`. The default is `true`. If you allow it to `useColor`, it will try to detect whether the user's monitor has color and use it if available. If the user's monitor does not have color and this can't be correctly detected, some of the displays in gray scale may be very hard to read.

- `readErrlog`
  You can set this option to either `true` or `false`. The default is `true`. If you run more than one diagapplet connected to the same controller, or another program that reads the error log, you may want to set this to false so that only one program will display error messages.

- `debug`
  This can be set to either `true` or `false`. The default is `false`. If it is `true`, it turns on all the `debug_flags` and causes a large amount of info to be printed to the Java console.

- `SourceCodeDirectory`
  The URL of an extra directory to look for source code in.

- `ListModulesByNumber`
  This can be set to either `true` or `false`. The default is `false`. The modules are normally listed alphabetically, but they can be listed instead by the module numbers in the hierarchy file.

As you can see, these parameters correspond to clicking some check boxes or specifying some paths or files after the applet is started. To see this, compare this with the discussion in the preceding section. Using these parameters allow us to start the applet in a known initial state. Note also that some of the parameters have default values which will be assumed automatically if you do not set any value for them. Example 9.2 provides a simple illustration for an HTML file for the RCS diagnostics tool.

Example 9.2: Example HTML file for the RCS diagnostics tool

```
<html>
<head><title>RCS Diagnostics Tool</title></head>

<body>
<h1> RCS Diagnostics Tool </h1>
<hr>

<applet
    codebase="/usr/local/rcslib/plat/java/lib/"
    code="diagapplet.diagapplet.class"
    id=diagapplet
    width=900
    height=600>

    <param name=hierarchyFile value="/home/veys/ivhs/ivhs.cfg">
    <param name=refreshTime value=0.25f>
    <param name=useColor value=false>
    <param name=connectOnStartup value=false>
    <param name=SourceCodeDirectory value="/home/veys/ivhs/src">
</applet>

<hr>
</body>
</html>
```

The diagnostics tool provides a high-level interface to RCS controllers running either on the same computer or a remote terminal. Thus, the diagnostics tool need not be located near the running controllers, since an NML server can connect the diagnostics with the running RCS modules. This allows for remote control over the RCS controllers.

In an RCS application, no matter how distributed it is, it is possible for a human operator to use diagnostics on a single terminal that is physically far away from the plant. You can send high-level commands such as "Start Operation," "Stop Operation," "Set Objective" (e.g., reference point, chemical concentration, etc.), "Switch Control Algorithm," and so on. The operator can make these decisions based on objectives of the control system and the data observed using the diagnostics tool. This allows the knowledge and skills of humans to be exploited at a high level in interacting with intelligent and (partially) autonomous control systems.

Chapter 10

Code Generation and Design Tool

In the preceding chapters of this book we introduced the basic building blocks of the RCS library and showed the reader how they can be used for hierarchical and possibly distributed control system development. We explained the basic parts of the library such as NML and programming in NML, defining messages for sharing information, implementing control nodes or modules, and so on. The reader should have noticed that most of the RCS code development is straightforward and systematic. Moreover, the development of some of the code is even application independent. For example, assume that the programmer defined a message called SYSTEM_INIT; then the following statement is needed:

```
#define SYSTEM_INIT_TYPE 201
```

in the same header file and the statement

```
case SYSTEM_INIT_TYPE :
     ((SYSTEM_INIT *) buffer)->update(cms);
     break;
```

(or similar statement depending on the definitions of the variables, such as buffer) in the format function. Similar arguments hold for the other parts of the program. Most application-dependent parts of the code development are in the implementation of the state tables or the algorithms for performing a particular command or operation.

This suggests the idea that it is possible to automate the generation of at least some of the RCS code in order to reduce the coding overhead of the programmer. Moreover, by standardizing the directory structure of RCS applications, it is possible to develop a tool to create a skeleton of any RCS application and to generate even the makefiles for compiling and running the application.

Based on these ideas, two new tools of the RCS library were developed by NIST. These are the NML code generator and the RCS design tool. The NML code generator is a tool which can automatically generate C++ and Java code for an RCS application. The RCS design tool, on the other hand, is a graphical program that helps you to "design" an RCS application visually. It includes the code generation utility in itself; therefore, with the design tool you can create most of the code, including configuration files, architecture files, makefiles, C++ code for header files, NML servers and so on, needed to construct and run an RCS application. In this chapter we discuss these utilities in detail.

10.1 NML Code Generator

The NML code generator is an important feature of the RCS Library. Using it you can generate both C++ and Java source code for an RCS application. It takes the definitions of NML messages from C++ header files and can create the following:

- C++ update functions, which call CMS update functions for updating the variable in the messages.
- The C++ format functions, which decide which update function to call based on the type of message received.
- C++ constructors for these messages.
- Java message classes.
- The Java message dictionary.

It is a Java program that can be run in four ways:

1. As an applet.
2. As an interactive stand-alone graphical program.
3. As a component of the RCS design tool.
4. As a noninteractive program suitable for use in a makefile.

Note that the first three options of the code generator are interactive. In other words, you have a graphical display as shown in Figure 10.1, where you can see the message definitions, the code corresponding to which can be created.

10.1.1 Interactive Operation

Use of the NML code generator as an interactive program is very easy and is described briefly here. For the following steps, refer to Figure 10.1:

1. Enter the path and URL or file name of the C++ header file or diagnostics configuration or architecture file into the text field at the top of the applet. You can use the "BROWSE" button to search for the file if you do not know where the file is located.

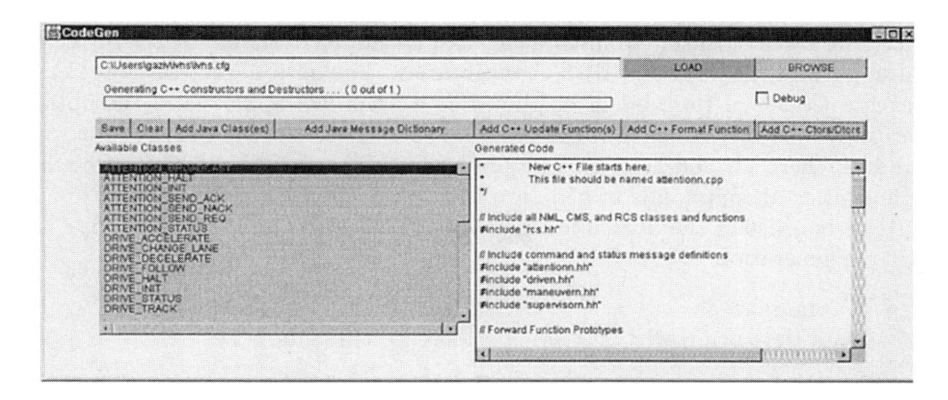

Figure 10.1: NML code generator.

2. Select NML messages for which update and/or format functions should be created from the "Available Classes" that were found and are listed on the left.

3. Press one or more of the "Add" buttons to create the source code. The code created appears in the "Generated Code" window.

4. Press the "Save" button to specify the name of the file to save the generated code to, or copy and paste the code into a text editor.

Clicking the `Debug` checkbox will turn the debugging on or off. It is used for printing debug messages.

As you can see, it is very easy to use the code generation utility of the RCS library; however, you should be aware of the following issues:

- Once any C++ code has been generated, the tool will not allow creation of Java code, or vice versa, until the "CLEAR" button is pressed.

- Some browsers do not allow Java applets to read or write to local files. Microsoft Internet Explorer 3+ and Netscape 4+ have the same restriction unless these privileges are granted during loading of the applet.

10.1.2 Execution from Within a Makefile

In order to be able to run the code generation tool noninteractively, you need to have the Java Developer's Kit installed on your system. Moreover, the RCS Java libraries should be in the `CLASSPATH`. Then you can run a command similar to the following:

```
java -classpath $CLASSPATH diagapplet.CodeGen.CodeGen
display_on=false script=yourscript.gen
```

where the value of the enviromental variable `$CLASSPATH` should be set correctly to the path of the Java and RCS Java libraries. The option `display_on=false` prevents CodeGen from being displayed as a graphical applet, and the option `script=yourscript.gen` tells the program to open up the file `yourscript.gen` and look there for commands. The file `yourscript.gen` should be a text file with a series of commands to generate the C++ or Java files.

Here is a list of the available commands that can be used in the script file for code generation:

- `# Comment`
  Lines that begin with a `#` are considered comments.

- `debug` *on/off*
  Turn on or off the printing of debug messages, useful for finding incompatibilities in your C++ header files.

- `load` *filename*
  Read the C++ header file, *filename*, for NML message definitions.

- `select_from_file` *filename*
  Select the classes from the given file for use by later commands.

- `cd` *directory*
  Change to the given directory.

- `set_format_function` *format_function_name*
  Set the name of the format function to generate.

- `package` *java_package_name*
  Set the name of a Java package to put any Java classes into.

- `generate java classes > *`
  Create Java classes from each of the currently selected C++ classes and place them in separate files.

- `generate java dict >` *filename*
  Create an NML message dictionary in Java using the currently selected classes and put it in the given file.

- `generate C++ update >` *filename*
  Create C++ update functions for each of the currently selected classes and append them to the given file.

- `generate C++ format >` *filename*
  Create a C++ format function using the currently selected classes and append it to the given file.

- `generate C++ constructor >` *filename*
  Create C++ constructors for each of the currently selected classes and append them to the given file.

- `clear`
  Deselect all the classes.

- `exit`
  Causes the code generator to exit, ignoring any commands after this.

In Example 10.1 we illustrate its use in an example which can be used for creating the format function, the update functions, and constructors for the messages of the drive module of an intelligent vehicle.

Example 10.1: Example code generation script

```
# CodeGen Script file to create driven.cc from driven.hh
load driven.hh
clear
select_from_file driven.hh
set_format_function driveFormat
generate C++ format >driven.cc
clear
select_from_file driven.hh
generate C++ update >driven.cc
generate C++ constructor >driven.cc
exit
```

10.1.3 Execution as an Applet

In order to view the RCS code generation tool as an applet we can use any Web browser which supports Java. If you have the Java Developer's Kit installed on your system, then another option would be to view it using the Java `appletviewer` program or to run it as a stand-alone application. If we use a browser or `appletviewer`, we need to prepare an HTML file for the tool. Within the HTML file it is possible to set the initial values of the code generation tool parameters. The available parameters are the following:

- `HFile, HHFile, HeaderFile`
  All mean the same thing and it only makes sense to specify one of them. It is the URL of a C++ header file to be loaded automatically and parsed just after the applet loads. You can also specify the name of an architecture file for this option.

  It might be worth noting that the URL should be on the Web server the applet is loaded from, or the browser may prevent the applet from loading the file. We recomend that the reader use `HFile` for strictly C headers, `HHFile` for C++ headers, and not use `HeaderFile` at all.

- `debug_on`
  This can be set to either `true` or `false`. The default is `false`. If set to `true`, it has the same effect as clicking the debug checkbox, which means the applet will print most of the debugging of information to the Java console about how the applet is parsing files and generating code.

- `UseDefaultTypes`
  This can be set to either `true` or `false`. The default is `true`. The code generator loads some information about common classes used in NML, including `NMLmsg`, `RCS_CMD_MSG`, `RCS_STAT_MSG`, and most of the Posemath classes. If you want to provide an alternative definition of one of those classes, you could set this to `false` and then have CodeGen load a file with your alternative definition.

- `script`
  The URL of a CodeGen script to be loaded automatically.

Example 10.2 provides a simple illustration for an HTML file. Note that in the `HHFile` option we have specified an architecture file instead of a header file. This will result in the loading of all the header files specified in the arthitecture file.

Example 10.2: Example HTML file for the code generation tool

```
<html>
<head>
<title>NML Code Generator</title>
</head>

<body>
<h1> NML Code Generator</h1>
<hr>

<applet
    codebase="/usr/local/rcslib/plat/java/lib/"
    code="diagapplet.CodeGen.CodeGen.class"
    id=CodeGen
    width=900
    height=500>

 <param name=HHFile value="/home/veys/ivhs/ivhs.cfg">
 <param name=UseDefaultTypes value=false>
 <param name=debug_on value=false>
</applet>

<hr>
</body>
</html>
```

To view the code generation tool using `appletviewer`, you can simply type

```
appletviewer CodeGen.html
```

on the command line from the directory where the HTML file is located. On the other hand, to run it as a stand-alone application we can simply type

```
java -classpath $CLASSPATH diagapplet.CodeGen.CodeGen
```

on the command line. We also need the `PATH` variable to be set correctly to the directory where `appletviewer` and `java` executables are located.

10.2 RCS Design Tool

The RCS design tool is a Java-based graphical program that allows users to lay out or build a hierarchy of controllers in a complex control system and generate automatically the C++ source code, makefiles, and configuration files needed to construct and run an application. Using this program, the user can easily add or remove new modules to the application, can modify the hierarchy, set up communication channels between modules, define messages to be passed between the modules, and so on.

On the top of the screen of the design tool there are buttons for creating the source code, running the makefile interpreter, running the application, and printing the graphical view of the hierarchy. A progress bar shows the progress and name of major tasks, such as creating source, which may take a few seconds.

Next, we discuss the generated files and conventions and the graphical controls in more detail.

10.2.1 Generated Files

The RCS design tool generates several different types of files needed by an RCS application. The detailed information needed to understand and work with each type of file was provided earlier in this book. Here, we overview them briefly.

NML message header files are C++ header files that define the classes of objects that can be sent to the module as commands or received from the module as status. Refer to Chapter 5 for more information on message classes. The files are named according to the following convention:

```
<module name>+'n.'+<C++ header ext.>.
```

For example, for a module called `maneuver`, the header file generated will be `maneuvern.h`. (Some programmers prefer `.h` as a C++ header extension, while others use `.hh` or `.hpp` to distinguish them from C header files; this is an option in the design tool.) The design tool generates a class for every command entered and one class for status. The programmer needs to add the needed field variables to these classes. For example, consider the `DRIVE_CHANGE_LANE` command for the drive module. This command should pass the required lane (either the lane to the left or right of the vehicle) to move to. Therefore, the programmer needs to add a variable, for example, `ref_lane`, to the message structure for holding the value of the required lane.

Module header files define the C++ class that contains all the functions for the module. (The RCS control module was discussed in Chapter 6.) These functions are called by the `NML_MODULE` base class when the appropriate message is received, or every cycle, but are not used by other modules directly.

Module source files provide "stubs" for each command that the module accepts. Inside these stubs, the programmer is expected to add state tables related to each command and probably send commands to its subordinates using NML communication channels that have already been set up. The state tables for `INIT` and `HALT` commands are already set up by the design tool. The code from the state table can be made available to the RCS diagnostics tool so that the current line in the state table can be watched.

Main source files contain the `main` function used to start execution. The `main` function creates an object for each type of module. Each module's controller function is called inside a loop that pauses for a timer each cycle to provide consistent cycle times. Other files generated include the following:

- NML configuration files that allow communications parameters and protocols to be selected.
- Makefiles that control how the source code is compiled.
- A configuration or architecture file used by the RCS diagnostics tool.
- A shell script that allows multiple executables to be started with a single command.

Using the RCS design tool for automatic code generation is very convenient; however, there are some issues that the RCS programmer should be aware of. These issues are the facts that the design tool generates a particular directory structure and all the makefiles and configuration files are generated based on this directory structure. Moreover, there are some architecture file attributes generated by the design tool which are not used by the diagnostics tool. Furthermore, the design tool uses some conventions in generating some of the C++ header and source files. In the following section, we describe the directory structure generated by the design tool for RCS applications. It is important because the makefiles for compiling and running the application are generated based on this directory structure.

10.2.2 RCS Directory Structure

The RCS design tool assumes that, in general, an RCS application will be developed by more than one programmer simultaneously. Therefore, each application is expected to have one main application directory and to mirror the directory tree under this directory for each programmer working on the application. This allows each programmer to change the source code, recompile, and test those changes in their own private workspace before releasing the changes to be used by the rest of the programmers. This is obtained by using some type of a *version control system* (VCS), which allows multiple users to share files and to edit the same file. It contains multiple revisions of the files (edited possibly by different users), a list of users that edit the file, and some other attributes. When a file is checked out and locked by a user, it cannot be edited by another user unless it is released by the first user. This prevents two (or more) users from editing the file simultaneously (i.e., it implements a type of mutual exclusion).

The main application release directory should contain the directory tree shown in Table 10.1, where the directory name `app` is a meaningful application-specific name. For example, for the "intelligent vehicle" discussed in this book, it will be `ivhs` if we supply this name to the design tool. The directory `src` is the

Table 10.1: RCS directory structure.

<table>
<tr><td rowspan="30">app</td><td colspan="3">Makefile</td></tr>
<tr><td colspan="3">Makefile.inc</td></tr>
<tr><td colspan="3">app.cfg</td></tr>
<tr><td colspan="3">app.nml</td></tr>
<tr><td colspan="3">run.app</td></tr>
<tr><td colspan="3">VCDIR</td></tr>
<tr><td rowspan="15">src</td><td rowspan="3">main</td><td>Makefile</td></tr>
<tr><td>main function files</td></tr>
<tr><td>VCDIR</td></tr>
<tr><td rowspan="3">intf</td><td>Makefile</td></tr>
<tr><td>NML message files and scripts</td></tr>
<tr><td>VCDIR</td></tr>
<tr><td rowspan="2">util</td><td>Makefile</td></tr>
<tr><td>VCDIR</td></tr>
<tr><td rowspan="3">module1_subdir</td><td>Makefile</td></tr>
<tr><td>module1 source and header files</td></tr>
<tr><td>VCDIR</td></tr>
<tr><td rowspan="3">module2_subdir</td><td>Makefile</td></tr>
<tr><td>module2 source and header files</td></tr>
<tr><td>VCDIR</td></tr>
<tr><td colspan="2">⋮</td></tr>
<tr><td rowspan="9">plat</td><td rowspan="4">linux</td><td>bin</td></tr>
<tr><td>include</td></tr>
<tr><td>lib</td></tr>
<tr><td>src</td></tr>
<tr><td colspan="2">⋮</td></tr>
<tr><td rowspan="4">win32msc</td><td>bin</td></tr>
<tr><td>include</td></tr>
<tr><td>lib</td></tr>
<tr><td>src</td></tr>
</table>

main source directory for all the code. It contains different subdirectories which contain the main functions, the NML message and script files, utility programs, and a directory for each module in the application. For the "intelligent vehicle" these directories would be `supervisor`, `maneuver`, `attention`, and `drive`. The VCDIR in each directory is a link to the release directory the version files are in. The `Makefile` under `app` is a top-level makefile, and the `Makefile.inc` is an application-specific include makefile. Moreover, each subdirectory has its own makefile for compiling the code within it (see Appendix B for more information on RCS makefiles).

The name `plat` stands for platform and has a tree of subdirectories for each platform that the application is developed for. Each platform specifies a particular compiler, CPU type, operating system, and sometimes certain compiler options, such as whether to compile a Windows program as a 16- or 32-bit application. For each platform there will be a subdirectory in both the programmer's workspace directory and in the application release directory. The names of the subdirectories can be `linux`, `sunos4`, `sunos5`, `irix5`, `dos_bor`, `dos_msc`, and so on. (See the table of tested platforms in Appendix E.) The directory of each specific platform contains four subdirectories, `bin`, `include`, `lib`, and `src`. These are the directories where the code is placed after compilation of that application. In other words, during compile time the source code files are copied to the `src` and the header files to the `include`. Then the object files are created in the `lib` directory and the executable is placed in `bin`, and the application is ready to run.

If you are not using the RCS design tool, you need not adhere to this directory structure. In that case, you should develop your own scripts for compiling and linking the application.

Having discussed the files and directory structure generated by the RCS design tool, in the rest of this chapter we discuss how you can use it to design an RCS application.

10.2.3 Graphical Controls

This section lists the various controls or views found in the design tool as well as their purpose. The controls are divided into five sections: *Top Buttons, Hierarchy view, Options view, Loops/Servers view, NML Code Generation view*, and *Files view*. The buttons on the top are always available no matter which view is selected, whereas the others are available only when a particular view is selected from the drop-down list near the top left corner.

Top Buttons

A row of buttons on the top is available regardless of the view selected. These buttons are the following.

- `Create Source`
  Creates all the header and source files, configuration files, architecture files, makefiles, and the directory structure of the application.

- `Print`
  Prints the graphic of the hierarchy.

- `Make`
  Runs the makefile interpreter in order to compile the application. Implements the command specified in the `Make Command` box of the `Options view`.

- `Run`
  Runs the application and (possibly) the diagnostics tool to control it. Im-

plements the command specified in the `Run Command` box of the `Options view`.

- `Import`
  Imports another controller into this application. A dialog window will open for the user to select the architecture or `.cfg` file of the controller to import. The top-level module of the selected controller will be added as a subordinate of the module currently selected.

- `Open`
  Opens the architecture or `.cfg` file of a controller to view or modify.

- `New`
  Begins working on a new RCS application.

- `Diag`
  Runs the RCS diagnostics tool for this application. Implements the command specified in the `Diag Command` box of the `Options view`.

- `Help`
  Opens a Web browser and connect to a Web site that contains information on the RCS library.

Note that you should have specified the options of the application, to be discussed later, correctly so that operation of buttons such as `Make`, `Run`, and `Diag` proceeds successfully.

Hierarchy View

When this option is selected, the appearance of the screen becomes as shown in Figure 10.2. The left side has a set of controls used to add modules, select subordinates, and add commands and auxiliary channels. The right side has a graphical depiction of the controller's hierarchy. Subordinates are placed below their supervisor with a line connecting them. Supervisors are modules which send commands to their subordinates and monitor the subordinates' status. The currently selected module will appear highlighted.

The module can be selected by clicking either on its icon or its name in the modules list. To add an additional module, enter the name in the text field under "Add Module." It will be a subordinate of the currently selected module. A module can be deleted by selecting it and clicking the "Delete Module" button.

Under the "Delete Module" button there is a drop-down list that allows the user to select what information is displayed about the current module. This box currently shows `Commands` in Figure 10.2. There are the following choices:

- `Subordinates`
  This option provides a list of subordinates. If a module from the list of the modules is selected, its subordinates on the list of subordinates are highlighted. The hierarchy can be reorganized by selecting and deselecting subordinates in the list.

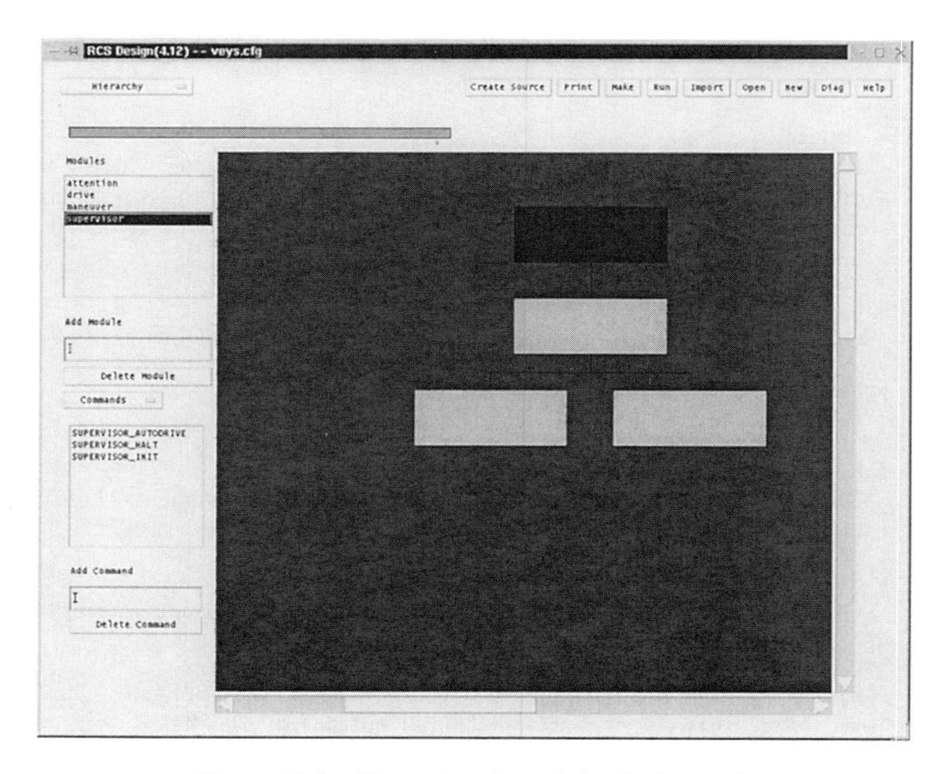

Figure 10.2: Hierarchy view of the design tool.

- `Commands`
  This option provides a list of commands that the selected module will accept. For example, in Figure 10.2 the module called **`supervisor`** accepts three commands, which are **`SUPERVISOR_INIT`**, **`SUPERVISOR_HALT`**, and **`SUPERVISOR_AUTODRIVE`**. Commands can be added to the list by typing them to the "Add Command" box and can be deleted from the list by selecting them and pressing the "Delete Command" button.

- `Aux. Channels`
  This option provides a list of auxiliary channels. Auxiliary channels provide a general method of communication that can be used between any two modules regardless of where they are in the hierarchy. You can select any module and add or delete an auxiliary channel for it. The auxiliary channels can be either input or output channels based on the information flow in the channel. You need to add the same auxiliary channel to both of the modules that will communicate through it. After the channels are set, a line is drawn between the two modules as shown in Figure 10.3. There, you can see that the supervisor has access to the channel called `attn_to_sup`, which is set as an input channel to the supervisor and will be updated each cycle.

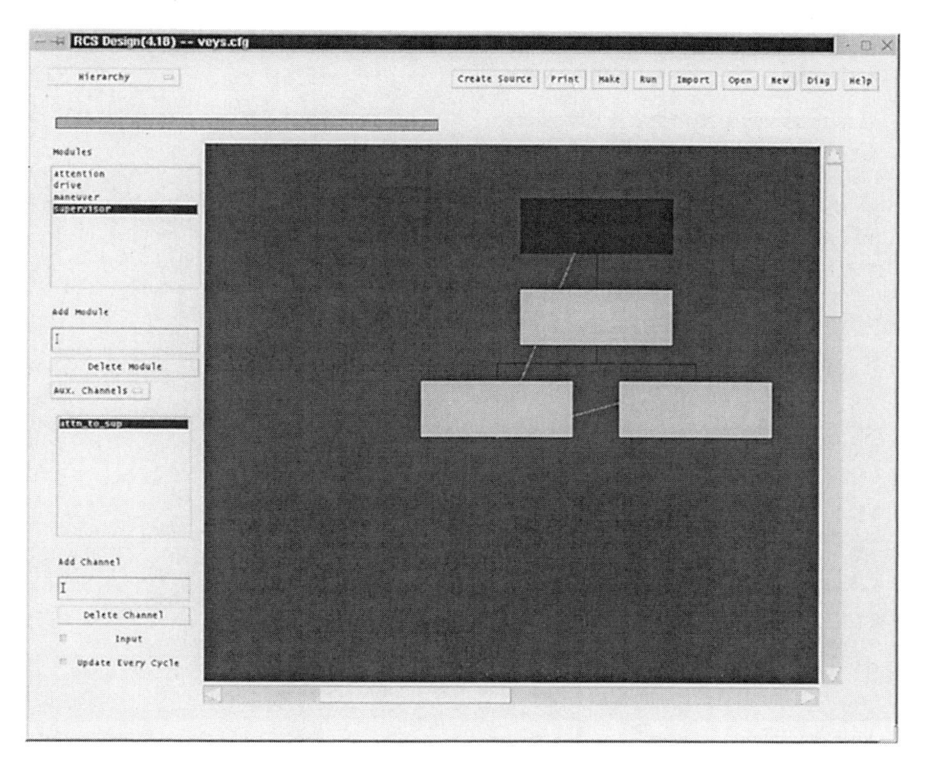

Figure 10.3: Auxiliary channels in the hierarchy view of the design tool.

Options View

The appearance of the tool when this view is selected is as shown in Figure 10.4. This view is populated with controls to change miscellaneous options such as the directories where files are written. The following is a list of options that can be controlled from this view (many of these options can also be set on the command line when the RCS design tool is run as a stand-alone application):

- `Application Directory`
  This is the directory where the release version of the application should be placed. For large RCS projects, with more than one developer, the application has a main application directory, and each programmer working on the application has a mirror of this application directory. For small projects, with only one programmer, this directory can be the same as the user directory.

- `User Directory`
  This is the directory where the development version for this user should be placed. In this private workspace the programmer can change the source

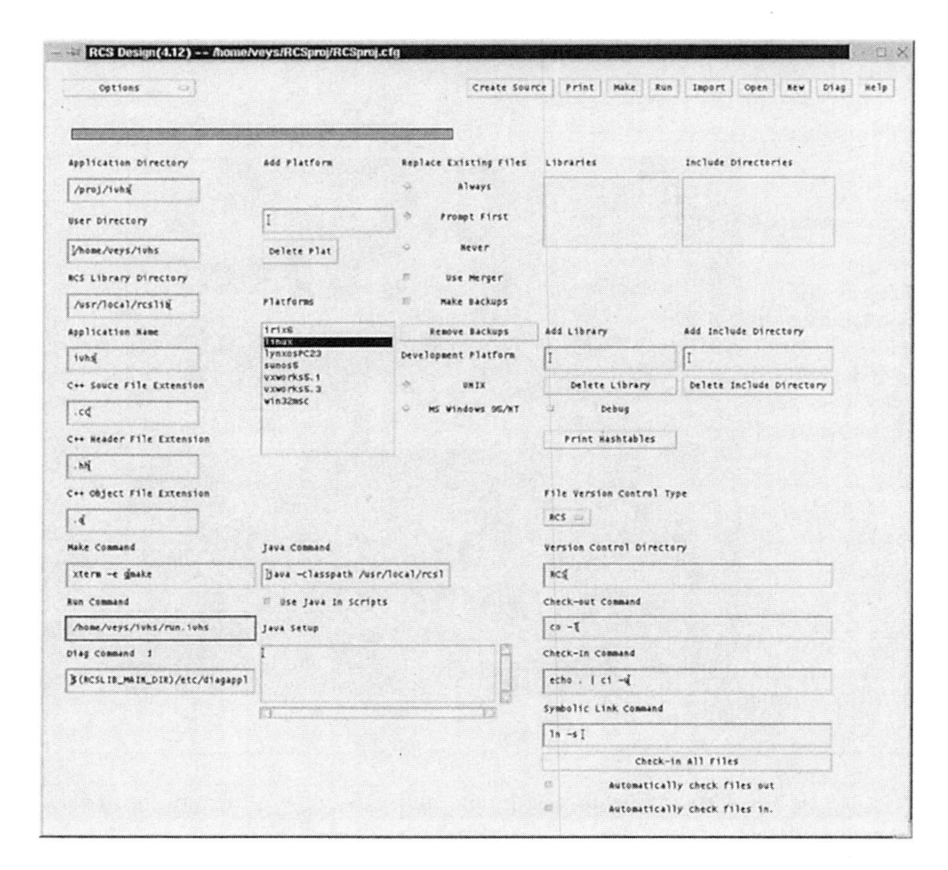

Figure 10.4: Options view of the design tool.

code, recompile, and test the changes made before releasing the code for the others.

- `RCS Library Directory`
  This box shows the path of the directory where the RCS library is installed. This is needed for compiling and linking the application properly.

- `Application Name`
  This is a name that is used in several source files to keep the files in this application unique. It is recommended that the name contain no spaces or punctuation. It must be less than or equal to five characters to work in DOS.

- `C++ Source File Extension`
  This box allows the user to chose the extension to use for the C++ source files (i.e., `.cpp`, `.C`, `.cc`). Generally, Microsoft Windows users use `.cpp` and UNIX users use `.cc`.

- `C++ Header File Extension`
  This is a box for choosing the extension to use for C++ header files (i.e., `.hpp, .h, .hh`). Generally, Microsoft Windows users use `.hpp` and UNIX users use `.hh`.

- `C++ Object File Extension`
  This is the extension that your C++ compiler adds to object files (i.e., `.o, .obj`). Generally, Microsoft Windows users use `.obj` and UNIX users use `.o`.

- `Make Command`
  This box supplies a command that, when executed, will invoke the makefile interpreter. There are some scripts included in the RCS library to solve some problems that occur on some systems when the make interpreter is invoked directly.

- `Run Command`
  On this box the user can supply a command to execute to run the application. A script file is normally generated automatically for this purpose.

- `Diag Command`
  On this box the user can specify a command that can be used to run the RCS diagnostics tool. For this purpose there is already a script provided under the `RCSLIB_MAIN_DIR/etc/` under the name `diagnostics` or `diagapplet`. `RCSLIB_MAIN_DIR` refers to the main directory where the RCS library is installed.

- `Platforms`
  The design tool can build an RCS application for different platforms. The user can add to or delete from the list of those platforms on which the application will be used. (See Appendix E for a table of platforms tested.) For each platform there is a directory in the directory structure for object files and a generic makefile (`.def` file) under `RCSLIB_MAIN_DIR/etc/` directory. This makefile specifies the compiler options, what compiler to use, and so on. The user can build the same application for multiple platforms simply by selecting different platforms before each "make."

- `Replace Existing File`
  When a file already exists and the user tries to create the file again, either though the "Files view" or with the "Create Source," the tool can replace it with an updated version, leave it alone, or ask the user. The options listed here allow the user to choose one of these.

- `Use Merger`
  If this checkbox is checked, then whenever the tool replaces a file it will do a line-by-line comparison of the new file with the old file and attempt to preserve user edits in the old file and merge the two together.

- `Make Backups`
  If this checkbox is checked, then when the tool replaces an existing file it will first rename the file according to the formula

  ```
  <original file name>+".~n~",
  ```

  where n is an integer that makes the backup unique. (This preserves your work but can create a large number of backup files; all the backups can be deleted by clicking the "Remove Backups" button.)

- `Development Platform`
  There are two options which can be selected here: UNIX and Windows 95/NT. If UNIX is selected, the makefiles generated are compatible with the GNU's `make` program; if Windows 95/NT is selected, the makefiles will be compatible with `NMAKE`, a program that comes with Microsoft Visual C++.

- `Libraries`
  This box shows the list the extra object libraries that should be linked into the application. The user can add new libraries to the list by writing the name of the library in the box labeled `Add Library`, and can delete a library from the list by choosing it and pressing `Delete Library` button.

- `Include Directories`
  This box shows the list of extra directories where the compiler should look for include files. As in the libraries case, the user can add and remove new directories from the list.

- `Debug`
  Checking this box will force the tool to print out a variety of messages that you can use for debugging the tool.

- `Java Command`
  The text from this field is used in scripts for starting the application. Its use is for the RCS diagnostics tool. In other words, the text from this box is inserted in the script file on the line starting the Java applet for the diagnostics tool.

- `Use Java in Scripts`
  If this checkbox is not checked, the RCS diagnostics tool will not be started in the script. In other words, there will not be an entry in the script file for starting the Java applet. Moreover, the options `Java Command` and `Java Setup` will not be used.

- `Java Setup`
  If the box `Use Java in Scripts` is checked, the lines defined in this box are inserted near the beginning of the script file. The three environment variables `PATH`, `CLASSPATH`, and `LD_LIBRARY_PATH` (on UNIX systems) need to be set up in this option.

- `File Version Control Type`
  This option is to specify the version control method to be used. The available options are *source code control system* (SCCS), *revision control system* (RCS), and NONE (no revision control is implemented). (Note: Please do not confuse the RCS for *real-time control system* with the RCS for *revision control system*. The RCS for *revision control system* is used only within this section.)

- `Version Control Directory`
  This option is to provide a name for the version control directories. The directories named with VCDIR on Table 10.1 will be created with the name provided here.

- `Check-out Command`
  This option is to provide the command to be used for checking out the files for editing. You should lock the files on checkout if you don't want other users to edit the same files simultaneously.

- `Check-in Command`
  This option is to provide the command to be used for checking in the files that were checked out so that the other users can have access to the changes that you have made. If files were locked on checkout, then they should be unlocked on checkin.

- `Symbolic Link Command`
  This option is to provide the command to be used for setting symbolic links on the platform that the design is carried on. This is used for creating symbolic links from the version control directory (VCDIR) of the user to the version control directory in the release directory, where the version files reside.

- `Automatically check files out`
  If this box is clicked, then the files of the application are checked out automatically before using or changing them. The `AutoCheckout` attribute in the architecture file is set to `true` (discussed later).

- `Automatically check files in`
  If this box is clicked, then the files of the application are checked in automatically when it is done with them. The `AutoCheckin` attribute in the architecture file is set to `true` (discussed later).

Loops/Servers View

If this option is selected, the appearance of the screen becomes as shown in Figure 10.5. The developer may use this view to assign modules to execution loops and to set up cycle times for these loops as well as for defining NML servers and assigning buffers to each server. As you can see from the figure, on the screen there are lists for the loops, the modules, NML servers, and buffers.

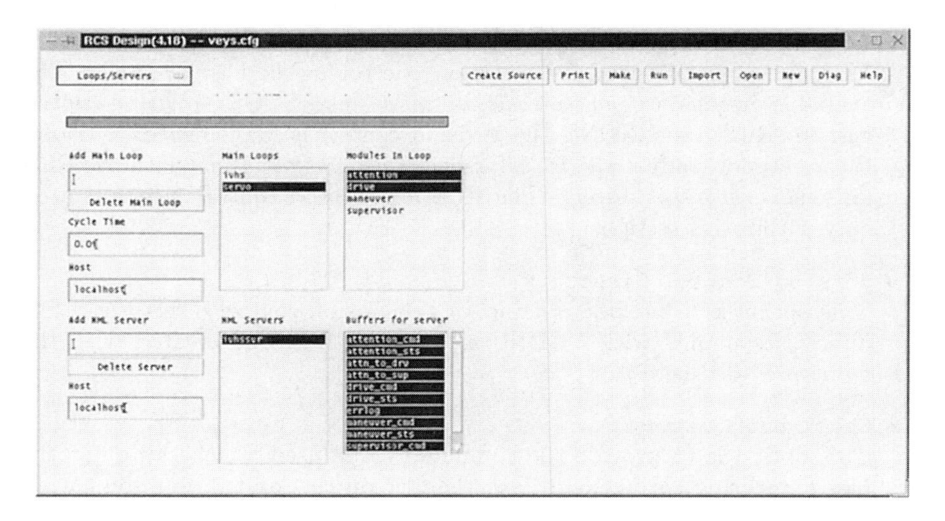

Figure 10.5: Loops view of the design tool.

A module can be assigned to only a single execution loop. This can be done by selecting the loop from the first list and then selecting the needed modules from the second. The loop selected and the corresponding modules will appear highlighted. For example, in Figure 10.5 the loop `servo` is selected and the `attention` and `drive` modules belong to this loop. On the other hand, although not shown, the `supervisor` and `maneuver` modules belong to the other loop, called `ivhs`. Different loops can be added by typing the name of the required loop in the "Add Main Loop" box, and can be removed by selecting them and clicking the "Delete Main Loop" button.

Each loop may have a different cycle time, which can be assigned by selecting the loop and typing the required cycle time in the "Cycle Time" box, which shows 0.01 in Figure 10.5 for the servo loop. All the modules in the same execution loop are linked together at compile time and must run on the same host at the same cycle time. We can specify the host on which the loop will be running by providing the name of the computer in the "Host" box. If the program will be running on the computer that the application is currently developed on, then one can simply specify `localhost` as a hostname. Assigning different modules to the same loop and therefore to the same executable may look like a drawback; however, there are advantages in synchronization to having the modules linked together.

Recall that we need to run NML servers for the buffers that will be accessed from remote hosts. The server needs to run on the same computer on which the buffer is located and will access it on behalf of the remote processes. Therefore, based on the physical location of the shared memory buffers in the application, we may need to run more than one NML server. You can add NML servers to the application simply by typing a name for the server in "Add NML Server" box. To delete a server, simply select it from the NML servers list and click the

"Delete Server" button. By typing the hostname on the "Host" box, you can specify the computer the server will run on. Choosing a server from the "NML Servers" list will highlight the buffers from the "Buffers for server" list that it will serve. By selecting and deselecting the buffers, you can assign or deassign particular buffers to the chosen server. Note that assigning a buffer to a server means that this buffer will be created on the host the server will run on.

NML Code Generation View

This view shows the NML code generator discussed in the preceding section. The appearance of part of the screen is as shown in Figure 10.1. The NML code generation tool is provided within the RCS design tool for convenience. It allows the user to generate C++ and Java format functions, update functions, and constructors for classes selected from C++ header file(s). The design tool automates the code generation by inserting scripts that run the tool from makefiles so that many users do not need to use the NML code generator directly. For more complete information about the code generation tool, see the preceding section of this chapter.

Files View

The files view allows you to see and update each of the files in the application. The view of the screen is as shown in Figure 10.6. You can choose a file from

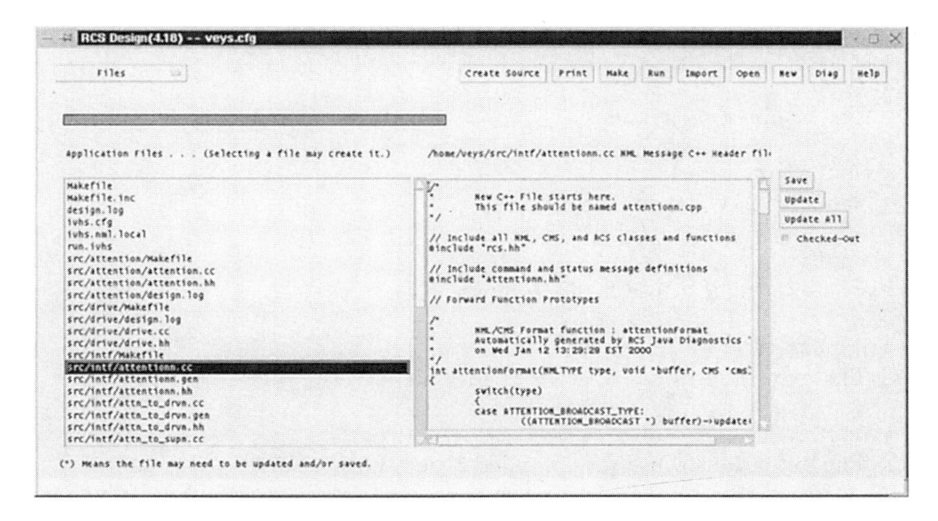

Figure 10.6: Files view of the design tool.

the box on the left, which lists all the files, by clicking on it. If the file does not exist or is not up-to-date when you select it from the list, it will be created and displayed. Otherwise, it will just be displayed on the simple text editor on the right. If you edit code in the text area, you can save your changes with the

"Save" button. To reintegrate changes from one of the other views, click the "Update" button. Pressing the "Update All" button will lead to the update of all the files in the application. The "Checked-Out" clickbox shows whether the file is checked out or not. By clicking on this clickbox you can check out or in the chosen file.

10.2.4 Design Tool as an Applet

As in the case of the NML code generator, to view the RCS design tool as an applet we need to prepare an HTML file for it. In this file it is possible to specify as parameters the options in the `Options` view of the tool. Below, we list these parameters with a short description. The reader may compare them with the ones described in the section on the `Options` view of the applet.

- `ConfigFile`
  The URL of a configuration or hierarchy file to load automatically.
- `UserDir`
  The starting value for the user directory.
- `AppDir`
  The starting value for the application directory.
- `RcsLibDir`
  The directory to look for the RCS library.
- `AppName`
  The application's name.
- `useColor`
  Same as useColor for diagapplet.
- `CppExt`
  The extension to use for C++ source files.
- `HppExt`
  The extension to use for C++ header files.
- `ObjExt`
  The extension to use for compiled object files.
- `MakeCommand`
  The command to call to build or make the application.
- `DevPlat`
  This can be set to either `MSWINDOWS` or `UNIX`. It sets the default value for a number of parameters to be more appropriate for either UNIX or Microsoft Windows.

- `JavaCmdPrefix`
  This is a string that is placed before the classname in scripts to start Java programs. Its default value is "`java -classpath $CLASSPATH`."

- `UseJavaInScripts`
  This can be set to either `true` or `false`. The default is `true`. If it is set to `false`, the design tool will generate scripts and makefiles that do not call for Java programs to be run.

- `JavaSetup`
  A string placed in scripts to set environment variables so that Java programs can be run.

- `JavaSetupFileName`
  The name of a file to be sourced in script files to enable Java programs to be run.

- `ImportDir`
  The directory to open the file dialog box with when the "Import" button is pressed.

- `UseMerger`
  This can be set to either `true` or `false`. The default is `true`. It sets the initial state of the checkbox on the options window for "Merger."

- `MakeBackups`
  This can be set to either `true` or `false`. The default is `true`. It sets the initial state of the checkbox on the options window for "Make Backups."

- `PLAT`
  Sets the platform initially selected.

- `CheckOutCommand`
  Sets the command used to check out files from SCCS or RCS (revision control system).

- `CheckInCommand`
  Sets the command used to check in files from SCCS or RCS.

- `SymbolicLinkCommand`
  Sets the command used to create symbolic links.

- `AutoCheckin`
  This can be set to either `true` or `false`. The default is `true`. Should the design tool automatically check in files to RCS or SCCS when it is done with them?

- `AutoCheckout`
  This can be set to either `true` or `false`. The default is `true`. Should the tool automatically check out files from RCS or SCCS before using or changing them?

- `ListModulesByNumber`
  Same as for diagapplet. In other words, this option can be either `true` or `false` and the default is `false`. The modules are normally listed alphabetically, but they can be listed by the module numbers.

- `VCT`
  "version control type" which must be either RCS, SCCS, or NONE.

Below we provide a simple ilustrative example for an HTML file for the RCS design tool.

Example 10.3: Example HTML file for the RCS design tool

```
<html>
<head>
<title>RCS Design Tool</title>
</head>

<body>
<h1> RCS Design Tool</h1>
<hr>

<applet
    codebase="/usr/local/rcslib/plat/java/lib/"
    code="rcsdesign.rcsDesign.class"
    id=rcsDesign
    width=1100
    height=700>

 <param name=ConfigFile value="/home/veys/ivhs/ivhs.cfg">
 <param name=RcsLibDir value="/usr/local/rcslib">
 <param name=UserDir value="/home/veys/ivhs">
 <param name=AppDir value="/proj/ivhs">
 <param name=PLAT value=linux>
</applet>

<hr>
</body>
</html>
```

As in the case of the NML code generator, you can view the design tool using a Web browser which supports Java or using the program `appletviewer`. To this end, you can load the HTML file within the browser or type

```
appletviewer rcsdesign.html
```

assuming that you have named your file `rcsdesign.html`.

It is possible to run the RCS design tool also as a stand-alone application. The command for this is similar to that for the NML code generator. In other words, you can just type

```
java -classpath $CLASSPATH rcsdesign.rcsDesign
```

on the command line.

This completes our discussion of the RCS design tool. It is a very useful tool that automatically generates most of the RCS application code. It writes the information about the controller structure into an architecture file that is used by the RCS diagnostics tool to connect to the controller. The information in the architecture file is used also by the RCS design tool itself while opening an existing application or importing one application controller into another. We discuss the architecture files in detail in Chapter 11.

Chapter 11

Architecture Files

An architecture file is a text file that provides information about the structure of the RCS controller, the command and status port numbers of the modules, buffer numbers of command and status buffers, auxiliary channels, NML servers, sampling rates, the paths and the names of the C++ header files containing the declaration of command and status messages, and so on. It can be created by the RCS design tool and is used by both the RCS design and diagnostics tools.

11.1 Modules

The architecture files consist of a series of *module blocks* (one for each module in the controller). Each module block has a *module name* followed by a series of *module attributes* separated by semicolons and surrounded by braces, as shown below.

```
MODULE_NAME{
    MODULE\_ATTRIBUTE=value;
    MODULE\_ATTRIBUTE=value;
    MODULE\_ATTRIBUTE=value;
}
```

Module attributes can be one of the following values:

- `child`
  This keyword specifies the child modules that this module supervises (i.e., the modules to which it sends commands to and/or reads the statuses of). Each module can have multiple children. Generally, the children are located one level lower in the RCS control hierarchy.

- `cmd_buffer_name`
  This attribute specifies the name of the buffer as defined in the NML configuration file which will be used to pass the command information to this module (see Chapter 7 for information on configuration files).

- `cmd_configuration_file`
  This attribute is for providing the name of the NML configuration file which specifies information such as the port number and buffer number for the command buffer.

- `cmd_port`
  The `cmd_port` is the TCP port where commands for this module should be sent. This number should be the same value as that used by the CMS/NML. Therefore, it should be taken from the NML configuration file. If the attribute `cmd_configuration_file` is specified, the data in the specified file will supersede this field.

  Note that the diagnostics tool does not work with UDP. For this reason, the user cannot specify the UDP port and is supposed to run an NML server using the TCP communication protocol for diagnostics to work.

- `cmd_buffer_number`
  The buffer number of the command buffer for this module. This value should be taken from the NML configuration file. If the attribute that is called `cmd_configuration_file` is specified, the data in the specified file will supersede this field.

- `cmd_types_file` or `cmd_types`
  This attribute specifies the path and the name of the C++ header file which contains the `NMLmsg` definitions for the commands that this module accepts.

- `stat_buffer_name`
  This attribute specifies the name of the buffer as defined in the NML configuration file which will be used to pass the status information of this module to the upper modules.

- `stat_configuration_file`
  This attribute is for providing the name of the NML configuration file which specifies the status port number and buffer number for the status buffer.

- `stat_port`
  Similar to the `cmd_port`, the `stat_port` is the TCP port where status for this module can be read. This value should be taken from the NML configuration file. If a `stat_configuration_file` is specified, the data in that file will supersede this field.

- `stat_buffer_number`
  This holds the buffer number of the status buffer for this module. This value should be taken from the NML configuration file. If the attribute that is called `stat_configuration_file` is specified, the data in that file will supersede this field.

- `stat_types_file` or `stat_types`
  This attribute specifies the path and the name of the C++ header file which contains the `NMLmsg` definitions for the status messages that this module may send out. If the definitions of the command and status messages are in the same file, then you can set the `cmd_types_file` attribute and omit this one.

- `predefined_types_file`
  Inside the command and/or status messages you may use predefined types such as `struct`, `class`, and `enum`, and the definition of these types may be provided in a separate file which can be included by using the `#include` directive. The diagnostic tool ignores `#include` directives within the header file, so it is necessary to include these C++ header files with the definition of predefined types in the architecture file. This attribute specifies the path and the name of such files. The next section is devoted to information on these files.

- `nml_configuration_file`
  If the status and command buffers are defined within the same configuration file, then instead of providing both the `cmd_configuration_file` and `stat_configuration_file` attributes, you may want to provide only one attribute. This attribute is for this purpose. Basically, it provides the name of the NML configuration file where the command and status port and buffer numbers are specified.

- `SourceCodeDirectory`
  Specifying this attribute is optional. Its purpose is to allow the user to specify a directory other than the "current directory" which contains the source code files of the application so that they can be displayed in the `State Table` view of the tool.

- `aux_input`
  This attribute specifies the name of auxiliary input channels that this module will use. The module can have many auxiliary input channels.

- `aux_input_types`
  This attribute specifies the path and the name of the C++ header file which contains the `NMLmsg` definitions for the auxiliary messages that this module may receive through this channel.

- `aux_output`
  This attribute specifies the name of auxiliary output channels that this module will use. The module can have many auxiliary output channels.

- `aux_output_types`
  This attribute specifies the path and the name of the C++ header file which contains the `NMLmsg` definitions for the auxiliary messages that this module may send out through this channel.

- `update_next_aux_every_cycle`
  This attribute can be either `true` or `false`. It is used to specify whether or not the information that will be passed through the auxiliary channels defined after it will be updated each cycle.

- `host`
  This attribute specifies the name of the computer that the module will run on.

- `class_name`
  This option specifies the C++ class name that will be used by the design tool when the module is imported into another application.

- `MainLoopName`
  In RCS applications there can be subsystems or loops which have common sampling times but which are different from those of the other loops. All the modules within a loop are combined within the same executable. This attribute provides the name of the loop to which this module belongs.

- `cycle_time`
  As stated above, the various loops, and therefore various modules, can have different cycle times in an RCS application. This option is for providing the length of the sampling time for this module.

- `release_library`
  The name and the path of the library files where the module will be stored. This option is used by the design tool when the module is imported into another application.

- `release_include_dir`
  The name and the path of the directory where the include files of the module will be stored. This option is used by the design tool when the module is imported into another application.

- `module_number`
  This attribute supplies a unique number for the module. It is used in a couple of places where the design tool needs a unique number. The `NMLTYPE` numbers for commands generated by the design tool, for instance, are computed according to the formula

  ```
  module\_number*1000 + command\_number.
  ```

These are basic attributes that are used in the module blocks in an architecture file for a given application. Some of these attributes are used only by the RCS diagnostics tool or by the RCS design tool, whereas some of them are used by both of the tools. Architecture files can be created or modified automatically by the RCS design tool. We provide them here because you need to understand them in case you want to understand and modify the code manually.

Example 11.1 provides an illustrative example for module blocks in an architecture file. The file shown is an excerpt from one possible architecture file for the modules of the hierarchy of an intelligent vehicle. Using the RCS diagnostics tool that will read the information about the controller structure from this file, the driver may watch the status of the vehicle and may decide, for instance, to switch to manual operation.

Example 11.1: Module blocks in an architecture file

```
attention{
        cmd_types="src/intf/attentionn.hh";
        stat_types="src/intf/attentionn.hh";
        SourceCodeDirectory="src/attention/";
        stat_buffer_name="attention_sts";
        cmd_buffer_name="attention_cmd";
        nml_configuration_file="ivhs.nml";
        class_name="ATTENTION_MODULE";
        MainLoopName="servo";
        cycle_time=0.01;
        host=localhost;
        update_next_aux_every_cycle=true;
        aux_output="attn_to_sup";
        aux_output_types="src/intf/attn_to_supn.hh";
        update_next_aux_every_cycle=true;
        aux_output="attn_to_drv";
        aux_output_types="src/intf/attn_to_drvn.hh";
        release_library="/proj/ivhs/plat/$(PLAT)/lib/libivhs.a";
        release_include_dir="/proj/ivhs/plat/$(PLAT)/include";
        module_number=3;
}

drive{
        ...
}

maneuver{
        ...
}

supervisor{
        child="maneuver";
        cmd_types="src/intf/supervisorn.hh";
        stat_types="src/intf/supervisorn.hh";
        SourceCodeDirectory="src/supervisor/";
        stat_buffer_name="supervisor_sts";
        cmd_buffer_name="supervisor_cmd";
        nml_configuration_file="ivhs.nml";
        class_name="SUPERVISOR_MODULE";
        MainLoopName="ivhs";
        cycle_time=0.1;
        host=localhost;
        update_next_aux_every_cycle=true;
        aux_input="attn_to_sup";
        aux_input_types="src/intf/attn_to_supn.hh";
        release_library="/proj/ivhs/plat/$(PLAT)/lib/libivhs.a";
```

```
        release_include_dir="/proj/ivhs/plat/$(PLAT)/include";
        module_number=1;
}
```

Note that some of the attributes listed above are omitted in this file. Such attributes are `cmd_port`, `cmd_buffer_number`, `cmd_configuration_file`, and so on. This is because the information specified in these attributes will be superseded either by information by another attribute or will be read automatically from the NML configuration file. In other words, specification of the `nml_configuration_file` is enough since it is a configuration file for both status and command buffers. Moreover, it contains all the port and buffer numbers of the shared memory buffers. We have not used the `predefined_types_file` attribute because no predefined classes are included in our header files.

11.2 NML Servers

The architecture file also holds information about the NML servers in the application, the hosts that the servers are running on, and the buffers that they are accessing. This information is used primarily by the RCS design tool for generating or modifying the C++ code for servers as well as the related scripts. Example 11.2 provides an example of server definition in an architecture file.

Example 11.2: NML server in an architecture file

```
ivhssvr{
        is_server=true;
        host="localhost";
        buf="attention_cmd";
        buf="attention_sts";
        buf="attn_to_drv";
        buf="attn_to_sup";
        buf="drive_cmd";
        buf="drive_sts";
        buf="errlog";
        buf="maneuver_cmd";
        buf="maneuver_sts";
        buf="supervisor_cmd";
        buf="supervisor_sts";
}
```

The `is_server` attribute is used to distinguish the server block from a module block in the architecture file. The `host` attribute provides the name of the computer the server will run on, and the `buf` attribute provides the name of the buffer the server will access (serve). A server may access more than one buffer.

The reader may recall the `Loops/Servers` view of the RCS design tool (refer to Figure 10.5) and make the correspondence between the information about the NML servers provided to the design tool there and the information in the architecture file. In fact, all the information about the controller carried in the application architecture file is somehow chosen by the designer via the RCS design tool.

11.3 Options

It is possible to specify (or override) in the architecture file some of the options or parameters set on the command line or in the HTML file for the RCS diagnostics or design tools. The format for specifying these options is as follows.

```
options{
    ATTRIBUTE=value;
    ATTRIBUTE=value;
    ATTRIBUTE=value;
}
```

Example 11.3 illustrates how we can specify these options in an architecture file. These options are useful in the case when the application is moved from one directory to another, or the directory for the RCS library or the file extensions used are not the default ones.

Example 11.3: Options in an architecture file

```
options{
        AppDir="/proj/ivhs";
        AppName="ivhs";
        CppExt=".cc";
        HppExt=".hh";
        ObjExt=".o";
        RcsLibDir="/usr/local/rcslib";
        ListModulesByNumber=false;
        MODULE_WIDTH=160;
        MODULE_HEIGHT=60;
        MODULE_X_SPACING=50;
        MODULE_Y_SPACING=40;
        VCT="RCS";
        AutoCheckin=true;
        AutoCheckout=true;
}
```

The attributes

`MODULE_WIDTH`, `MODULE_HEIGHT`, `MODULE_X_SPACING`, and `MODULE_Y_SPACING`

specify how the modules will be drawn on the graphical displays of the RCS design and diagnostics tools. The reader should be familiar with the other options from the earlier chapters on the RCS design and diagnostics tools.

This concludes the discussion of the architecture files for the RCS diagnostics and design tools. We want to emphasize once more that this file can be generated and modified automatically by the RCS design tool. The objective of the discussion here was to help the reader to better understand these options in case there is a need to modify them manually.

Part III

Appendices

Appendix A

C++ Introduction

It is assumed that users of the RCS library know the C and/or C++ programming languages. The purpose of this appendix is to teach C programmers just enough C++ so that they can use the RCS library.

A.1 Classes

Classes in C++ are user-defined data types (structures). Each class contains data and functions that manipulate the data. The data fields of a class are called *data members* and the function fields are called *function members*. Instances of a class are called *objects*. The keyword `class` is used to define data types (i.e., classes). (This is similar to the keyword `struct` in C.) The code in Example A.1 shows how a class is defined.

Example A.1: Definition of a class

```
class student {
public:        // Defines the access for the following members.
    char name[30];
    int age;
    float gpa;
    void get_name(char *s);
    void print_information();
};
```

Here, `student` is the name of the class; `name`, `age`, and `gpa` are the data members; and `print_information()` and `get_name(char *s)` are the function members of that class. Function members define operations that can be performed on this class and its members. If you declare a function member of a class, you should also implement this function. An example implementation of the function members of the `student` class is shown in Example A.2.

Example A.2: Possible implementation of the member functions of the class student

```
void student::get_name(char *s)
{
    strcpy(name,s);
}

void student::print_information()
{
    printf("Student name = %s, age = %d, GPA = %lf \n",
   name, age, gpa);
}
```

The keyword `public` is a *member access specifier* and it allows all the data members and function members of that class to be accessible by the program whenever the object itself is accessible by the program. The other two access specifiers, `private` and `protected`, are not considered here.

The user-defined classes are used in the program as any other data types such as `int`, `float`, and `char`. The code in Example A.3 shows how this class can be used.

Example A.3: Using user-defined classes

```
#include <studentClass.c>  // the file containing above code

main() {

    student std1;
    student std2;

    std1.age = 20;
    std1.gpa = 3.20;     // Accessing public data members
    std1.get_name("Kim Brown");

    std2.age = 21;
    std2.gpa = 3.54;
    std2.get_name("Joen Parkman");

    std1.print_information();   // Calling a public member function
    std2.print_information();
}
```

The output of this code is:

Student name = Kim Brown, age = 20, GPA = 3.200000
Student name = Joen Parkman, age = 21, GPA = 3.540000

A.2 Constructors

The class in the preceding examples has a problem: When an object of its type is created, its variables are uninitialized so its behavior is unpredictable. To avoid this situation, member functions called *constructors* are defined. Constructors are functions which, if necessary, take care of the initialization of the object created so that all new objects behave in a predictable manner. The name of the constructor should be the same name as that of the class itself. Classes can have zero, one, or more constructors. A constructor can take arguments, and in this case the arguments should be added to the declaration of any object of that class. A constructor is called automatically each time the object of that class is created (i.e., when the object is defined or the operator `new` is used).

In NML, constructors allocate memory, initialize communications channels, read configuration files, set the type and size of messages, and perform almost all of the run-time setup necessary for NML to work. In Example A.4 we redefine our class student and implement example constructors.

Example A.4: Possible constructor for the student class

```
class student {
public:        // Defines the access for the following members.
    student();
    student(char *s, int = 0, float = 0.0);
    char name[30];
    int age;
    float gpa;
    void get_name(char *s);
    void print_information();
};

student::student()           // Constructor with no arguments
{
    strcpy(name,"???");
    age = 0;
    gpa = 0.0;
}

student::student(char *st, int ag, float gp)
{   // This constructor can take 1,2 or 3 arguments
    strcpy(name,st);
    age = ag;
    gpa = gp;
}
```

Here, two separate constructors are defined. The first constructor is called if no arguments are specified in the definition, and the second is called if there are one or more arguments. The second and third arguments of the second constructor have default values, which will be assigned to the corresponding member if no value is specified. The code that shows how class objects can be defined is presented in Example A.5.

Example A.5: Defining objects of user-defined classes with constructors

```
#include "studentClass.c"

main() {

    student std1;                          // first constructor
    student std2("Nataly Kohno");          // second constructor
    student std3("Vladimir Drake",23);       // second constructor
    student std4("Abdul Guven",25,3.80);   // second constructor
    student *std_ptr;

    // here constructor is invoked
    std_ptr = new student("Adam Pirvomaysky",18,3.10);

    std1.print_information();
    std2.print_information();
    std3.print_information();
    std4.print_information();
    std_ptr->print_information();

    delete std_ptr;             // Calls the destructor
}
```

The output of this code is:

```
Student name = ???, age = 0, GPA = 0.000000
Student name = Nataly Kohno, age = 0, GPA = 0.000000
Student name = Vladimir Drake, age = 23, GPA = 0.000000
Student name = Abdul Guven, age = 25, GPA = 3.800000
Student name = Adam Pirvomaysky, age = 18, GPA = 3.100000
```

A.3 Destructors

A *destructor* is a special member function, which is the complement of the constructor member function. A destructor is defined with a *tilde* (~) in front of the class name. It is invoked automatically when the object is destroyed (i.e., when the program leaves the scope in which the object was defined or when the `delete` operator is used). Destructors are often used to deallocate memory, close communications channels and files, and to return the system to a known or stable state. The `student` class defined above does not need to have a defined destructor, since it is a very simple data structure. Therefore, you can define a destructor which has an empty body. For the sake of illustration, we define in Example A.6 a destructor which prints some information on the screen.

Example A.6: Destructor and its use

```
student::~student()        // Destructor
{
    printf("Destroying the object with field 'name' = %s \n", name);
}
```

The code shown next is to illustrate when the destructors are called.

```
#include "studentClass.c"

void ex_function() { // example function for illustration
    student std3("Tanya Starr",25);          // second constructor
    student std4("Peter Hopkins",25,3.80);  // second constructor
}

main() {

    student std1;                     // first constructor
    student std2("Babis Klein");      // second constructor
    student *std_ptr;

    // here constructor is invoked
    std_ptr = new student("Murat Kara",26,3.90);
    delete std_ptr;              // Calls a destructor
    ex_function();
}
```

The output of this program is:

```
Destroying the object with field 'name' = Murat Kara
Destroying the object with field 'name' = Peter Hopkins
Destroying the object with field 'name' = Tanya Starr
Destroying the object with field 'name' = Babis Klein
Destroying the object with field 'name' = ???
```

A.4 Function Overloading

In the C programming language the functions should have a unique name. In C++, however, different functions with the same name are allowed as long as they have a different set of parameters (arguments). This capability in C++ is called *function overloading*, and functions with the same name are called *overloaded functions*. When an overloaded function is called, the C++ compiler selects the proper function by examining the number, types, and order of the arguments in the call. Member functions of the classes can also be overloaded. We saw an overloaded constructor in the Example A.4. Another example is the function `square`shown in Example A.7. It is overloaded to accept `int` or `float` arguments and to return `int` or `float`.

Example A.7: Overloading a function to accept and return both `int` and `float`

```
#include <stdio.h>

int square(int x) { // func. for squaring integers
    return x*x;
}

float square(float x) {// func. for squaring floats
    return x*x;
}

main() {

    int x = 8;
    float y = 9.5432;

    printf("The square of integer  %d is %d \n",x,square(x));
    printf("The square of float  %lf is %lf \n",y,square(y));
}
```

This code produces the following output:

```
The square of integer  8 is 64
The square of float  9.543200 is 91.072655
```

Similar to function overloading in C++, you can also overload the operators such as (+), (-), (/), and so on, to perform different operations on user-defined classes. We do not consider these cases here.

A.5 Inheritance

Inheritance, which allows programmers to derive new classes from already existing ones, is an important feature of C++. The original class is called the *base class* and the new class is called the *derived class*. The derived class has all the members of the base class plus the new added members. Every object of the derived class is also an object of its base class. Inheritance allows reuse of previously developed software. Member functions of the base class defined as *virtual* can be redefined in the derived class (although they need not be) to perform different operations. This is done by declaring new functions in the derived class with the same name and arguments as those of the original function. If a new function with the same name is not defined, the function of the base class is used. Example A.8 shows the derivation of a class `gradStudent` from the class `student`.

Example A.8: Derivation of a class gradStudent from the class student

```
class gradStudent : public student
{
public:        // Defines the access for the following members.
    gradStudent();  // Constructor
    gradStudent(char *, int, float, char *, char *); // Constructor
    ~gradStudent();                // Destructor
    char *subject;
    char *title;
    void print_subjectAndTitle();
};
```

Because the derived class inherits the base class members, when a new object of the derived class object is created, the base class constructor should be called to initialize the base class members of the derived class object. A base class initializer can be provided in the derived class constructor; otherwise, the derived class's constructor will call the base class's default constructor implicitly. In the implementation shown in Example A.9, the first constructor of gradStudent calls implicitly the constructor of the class student, and the second constructor calls it explicitly.

Example A.9: Possible implementation of the member functions of the derived class gradStudent

```
gradStudent::gradStudent()
{
    printf("First constructor of gradStudent \n");
    subject = NULL;
    title = NULL;
}

gradStudent::gradStudent(char *st, int ag, float gp, char *sub, char *tit)
    : student(st,ag,gp)   // Call the base class constructor
{
    printf("Second constructor of gradStudent \n");
    subject = new char[strlen(sub)+1];
    strcpy(subject,sub);
    title = new char[strlen(tit)+1];
    strcpy(title,tit);
}

gradStudent::~gradStudent()
{   // memory allocated with new in the constructor should be deleted
    printf("Destructor of gradStudent \n");
    delete subject;
    delete title;
}

void gradStudent::print_subjectAndTitle()
{
    printf("Thesis Subject = %s, Title = %s  \n",subject,title);
}
```

In the section about *destructors* we indicated that the destructors could be used to free memory. Since we initialized the memory with the `new` operator in the constructor above, the destructor should free this memory, as shown in Example A.9.

Now we show how the derived class "inherits" (i.e., preserves) the data and function members of the base class. Consider the code shown in Example A.10 together with its output.

Example A.10: Illustrative use of the class gradStudent

```
#include "studentClass.c"

main() {

    student std("Tony Spenser",25,3.00);
    gradStudent gradStd1("Nikolay Zvezdev",26,3.80,
                         "Stability of Fuzzy Systems",
 "Research Assistant");
    gradStudent gradStd2;

    std.print_information();
    gradStd1.print_information();
    gradStd1.print_subjectAndTitle();
    gradStd2.print_information();
    gradStd2.print_subjectAndTitle();
}
```

The output of this code is:

```
Second constructor of gradStudent
First constructor of gradStudent
Student name = Tony Spenser, age = 25, GPA = 3.000000
Student name = Nikolay Zvezdev, age = 26, GPA = 3.800000
Thesis Subject = Stability of Fuzzy Systems, Title = Research Assistant
Student name = ???, age = 0, GPA = 0.000000
Thesis Subject = , Title =
Destructor of gradStudent
Destroying the object with field 'name' = ???
Destructor of gradStudent
Destroying the object with field 'name' = Nikolay Zvezdev
Destroying the object with field 'name' = Tony Spenser
```

As you can see, the class `gradStudent` and its constructors acted as they were defined in the way shown in Example A.11.

Example A.11: Illustrative use of the class gradStudent

```
class gradStudent :
{
public:                // Defines the access for the following members.
    gradStudent();
    gradStudent(char *, int, float, char *, char *);
```

```
    ~gradStudent();  // Destructor
    char name[30];
    int age;
    float gpa;
    void get_name(char *s);
    void print_information();
    char *subject;
    char *title;
    void print_subjectAndTitle();
};

gradStudent::gradStudent()
{
    printf("First constructor of gradStudent \n");
    strcpy(name,"???");
    age = 0;
    gpa = 0.0;
    subject = NULL;
    title = NULL;
}

gradStudent::gradStudent(char *st, int ag, float gp, char *sub, char *tit)
{
    printf("Second constructor of gradStudent \n");
    strcpy(name,st);
    age = ag;
    gpa = gp;
    subject = new char[strlen(sub)+1];
    strcpy(subject,sub);
    title = new char[strlen(tit)+1];
    strcpy(title,tit);
}
```

Note from the output of the program in Example A.10 that the destructors are called in the order opposite of that of the constructors.

A.6 Polymorphism and Virtual Functions

Polymorphism means "multiple shapes" and allows objects belonging to different classes to respond to the same message. Moreover, it allows sending a message without knowing the class of the object. This makes it easy to develop extremely general codes, which makes object-oriented programming useful. In C++ polymorphism is implemented via *virtual* functions. Virtual member functions are declared by placing the `virtual` keyword in front of the function definition. In the derived class you can redefine the virtual function from the base class so that the objects from the derived class use the function of the derived class. (Actually, every member function of the base class can be overwritten in the derived class, but nonvirtual functions do not allow for polymorphism.) This will be explained by an example. Let us redefine our class `student`, implement some of its member functions, and rederive the class `gradStudent` as shown in Example A.12.

Example A.12: Redefinition of the student class with virtual and nonvirtual functions

```
class student {
public:             // Defines access for the following members.
    student();      // Constructor
    student(char *, int = 0, float = 0.0);    // Constructor
    ~student();     // Destructor
    char name[30];
    int age;
    float gpa;
    void get_name(char *s);
    virtual void print_information_virtual();
    void print_information_nonvirtual();
};

void student::print_information_virtual()
{
    printf("BASE Virtual : Student name = %s, age = %d, GPA = %lf \n",
   name, age, gpa);
}
void student::print_information_nonvirtual()
{
    printf("BASE NONVirtual : Student name = %s, age = %d, GPA = %lf \n",
   name, age, gpa);
}
```

Note that we defined two functions for printing information to be able to compare the virtual and nonvirtual functions. Moreover, implementation of these functions is the same, for simplicity. The redireved class gradStudent and the implementation of its functions are shown in Example A.13.

Example A.13: The new gradStudent class

```
class gradStudent : public student
{
public:                // Defines the access for the following members.
    gradStudent();
    gradStudent(char *, int, float, char *, char *);
    ~gradStudent();  // Destructor
    char *subject;
    char *title;
    void print_information_virtual();
    void print_information_nonvirtual();
};

void gradStudent::print_information_virtual()
{
    printf("DERV Virtual : Student name = %s, age = %d, GPA = %lf \n",
   name, age, gpa);
    printf("Thesis Subject = %s, Title = %s \n",subject,title);
}
void gradStudent::print_information_nonvirtual()
{
    printf("DERV NONVirtual : Student name = %s, age = %d, GPA = %lf \n",
```

```
    name, age, gpa);
     printf("Thesis Subject = %s, Title = %s \n",subject,title);
}
```

Note that in the new declaration of this class we have two functions with the same name as in the base class.

The aim is now to show the differences between a virtual and a nonvirtual function when called through a pointer to a base class. Therefore, consider the code in Example A.14 and its output.

Example A.14: Difference between virtual and nonvirtual functions when called through a pointer to a base class

```
#include "studentClass1.c"  // Modified version with virtual functions

main() {

    gradStudent gradStd1("Cem Georgiev",26,3.50,
                         "Neural Networks","Graduate Fellow");
    gradStudent gradStd2("Halil Osman",30,3.40,
                         "Power Electronics","Teaching Assistant");
    gradStudent *gradStd_ptr;
    student *std_ptr;

    gradStd_ptr = &gradStd1;
    std_ptr = &gradStd2;

    gradStd_ptr->print_information_virtual();
    gradStd_ptr->print_information_nonvirtual();

    std_ptr->print_information_virtual();
    std_ptr->print_information_nonvirtual();
}
```

The output of this code is:

```
DERV Virtual : Student name = Cem Georgiev, age = 26, GPA = 3.500000
Thesis Subject = Neural Networks, Title = Graduate Fellow
DERV NONVirtual : Student name = Cem Georgiev, age = 26,GPA = 3.500000
Thesis Subject = Neural Networks, Title = Graduate Fellow
DERV Virtual : Student name = Halil Osman, age = 30, GPA = 3.400000
Thesis Subject = Power Electronics, Title = Teaching Assistant
BASE NONVirtual : Student name = Halil Osman, age = 30, GPA = 3.400000
```

Note that in Example A.14, `std_ptr` is a pointer to an object of class `student` but holds an address of an object of class `gradStudent`. In other words, in the last two calls in the program the object of class `gradStudent` was referenced through a pointer to its base class `student`. For this reason, there is a difference between these two calls. A call to `print_information_virtual()`, defined as a virtual function in the base class, led to calling of the function with the same name in the derived class, whereas call to the nonvirtual function `print_information_nonvirtual()` invoked the function of the base class, even the fact that there was a function with the same name in the derived class. Virtual functions are very important, because you can derive hundreds of new classes from a base class, and during programming you do not have to worry about which derived class an object belongs to, by using pointers to their base classes. For example, consider that we derived several other classes from the class `student`, such as `undergradStudent`, `professionalStudent`, and so on. Moreover, assume that every new class has its own `print_information()` function. If we declare this function as a `virtual` in the base class, we can deal with all the objects of these different classes by a pointer to a class `student`, say `std_ptr`. Then for any particular object from that class, which is derived or not, writing `std_ptr->print_information()` will call the appropriate function and we do not have to worry about which function would be called. This works only with *pointers* and *references*, because the objects of different classes have different sizes; however, pointers and references use addresses (which are always the same size).

Appendix B

Compilers and Makefiles

This is a short introduction to two of the tools that are needed to produce RCS applications—the C/C++ compilers and makefiles. The discussion here is centered around GNU's versions of the compiler and make, since these are widely available for multiple platforms (including UNIX, Linux, OS2, Windows 95/98/NT, etc.), and they are free.

B.1 gcc/g++ Compilers

GNU's C/C++ compiler, `gcc` and `g++`, can compile C, C++, or Objective C programs. Front ends for compiling other programs are under development. Currently, both Fortran and Pascal compilers are available. The compiler itself is integrated—simply typing `gcc` will compile all three of the C variations. Which language it uses to compile depends on the suffix of the source files. Files with the extensions `.cc`, `.C`, `.cpp`, and `.cxx` all default to the C++ compiler, even if you compile using the name `gcc`. Since many C++ programs require special class libraries, it is best if you explicitly use `g++` when compiling C++-specific code—`g++` is a program that calls `gcc` with the default language set to C++ [52].

The `g++` compiler is somewhat special as far as C++ compilers go, because it is what we could call a true-C++ compiler. That is, there is no intermediate transformation stage that C++-files go through before compiling. It takes C++ code and compiles directly into assembly code. Many current C++ compilers (even some good commercial ones) first transform the C++ code into equivalent C code and then run a C-compiler to create the appropriate assembly version; `g++` has no such intermediate stage.

Since the compilers are integrated, it is correct to use the term `gcc` when referring to either C or C++ programs. For the remainder of this section, we will always refer to it as `gcc` for simplicity, and the reader should realize that `g++` should be used when compiling C++ programs.

As can be seen, GNU's compiler is quite a powerful tool. `gcc` can even be installed as a cross-compiler, producing executable format for basically any platform, including Intel, Motorola, DEC, and so on, simply by choosing the correct command line options [52]. By default, gcc compiles code for the same type of machine that you are using. It also provides the means for producing shared libraries, static libraries, archives, and dynamic link libraries.

Because of its many abilities, gcc has a vast list of command line options. Basically anything you can do on a commercial C or C++ compiler (such as with bcc [Borland's C/C++ compiler] or with Microsoft Visual C/C++ [MSVC]) can be done with GNU's compiler. In this section we examine only the more popularly used options, such as those for suppressing warnings and changing optimizations. Many other options exist, of course, but these get quite complicated and you may not need most of them anyway.

B.1.1 Invoking gcc

The standard way to invoke gcc is simply to type at the command line

```
gcc filename1.c filename2.c ...
g++ filename1.cc filename2.cc ...
```

This will compile, assemble, and link the files filename*.c and produce an executable called a.out in the current directory; the user can then run this program simply by typing a.out at the command line. To assign a more meaningful name to the executable that is created, use the -o option, like this:

```
gcc -o runfile filename1.c filename2.c
```

This produces an executable called runfile.

When the programming becomes more involved, you will find that at times you do not want to produce a complete executable, but would rather simply have an assembly language version of your code, or perhaps instead, an object file. By default, when gcc is invoked, it typically does preprocessing, compilation, assembly, and linking. In the preprocessing stage the preprocessor statements in the code are compiled. Compilation then produces an assembly code version of the program. This assembly code is turned into machine code in the assembly stage, and linking resolves any external dependencies on other object files (and libraries) and produces an executable. With gcc, you can stop the compilation at any of the foregoing stages. For example, you can produce an object file (i.e., do everything but linking) using the -c option like this:

```
gcc -o filename.o -c filename.c
```

which produces the object file filename.o. This can be used later to link with other object files and produce an executable. In case you're interested, the summary of the options controlling the output of gcc follows:

-c Compile and/or assemble but do not link. Produces an object file. By default, gcc replaces the extension of the input file with a .o when producing the output filename.

-S Compiles the code but does not assemble. This produces an assembly code version of each input file. The default output filename is the input filename with a .s as the extension.

-E This option tells gcc to do preprocessing only, no compiling, assembling, or linking. Resulting output is sent to stdout (the screen).

When we get to the discussion on make, you will see that the -c is used quite frequently. The others are used only on rare occasions.

B.1.2 Special Command Line Arguments

Several other command line arguments are quite helpful in developing code and are used quite frequently. In particular, gcc offers several debugging options. These are used in conjunction with gdb, the GNU debugger, which takes quite a while to learn, so we will not discuss these options here (but it is helpful to know they exist; see the "man" page for more information). Here is a short list of command line options that are more popular:

-Wall This option exhibits all typical warning messages that are normally suppressed when running gcc. This is particularly helpful, because the warnings are often errors that are overlooked by the developer. With this option, warning messages such as

```
Unused local variable
Variable may be used before it is assigned
Comparing int to pointer
```

are displayed where normally, these warnings are suppressed.

-O, -O2, -O3 These are the general optimization options available in GNU gcc (there are other, more specific optimizations for both speed and executable size that are not discussed here). As you might have guessed, -O does a little bit of optimization, -O2 some more, and -O3 a whole lot. The more optimization you use, the longer it takes to compile, but the smaller and faster the code will be (of course, this is only *generally* true).

-D This option allows you to define a constant value much like the preprocessor command #define does. Often, this is used in debugging code, as this next example illustrates:

```
gcc -DDEBUG=1 filename.c
```

Here, `DEBUG` is assigned a value 1, and this value replaces any occurrence of `DEBUG` in any and all of the input files. Note that this is similar to placing

```
#define DEBUG 1
```

at the beginning of the input files.

`-I` This allows you to tell `gcc` explicitly where to look for header files. Often in programming development, you will have a specific directory for header files that is not in the default search path of `gcc`. If your header files are in the directory `/home/myname/include`, for example, then you can tell `gcc` this using the syntax

```
gcc -I/home/myname/include -o runfile -c filename.c
```

and `gcc` will search in this directory.

`-L` This is the same as `-I` except for libraries instead of header files.

`-l` Often, you will get an error such as, "`sqrt(): Symbol undefined.`" This occurs because the math library was not linked into object files. `gcc` only links in a few of the libraries by default, such as `stdlib`. For others you must link them in explicitly with the `-l` option. This is done as follows:

```
gcc -I/home/myname/include -o runfile filename.c -lname
```

This links in the library `libname.so`. In UNIX, libraries are usually named as shown here, starting with `lib` and ending with `.so` (shared libraries) or `.a` (static libraries). The meaningful part is the portion of the name in between (here we call it the root name). With this notation, libraries are specified in `gcc` in a shorthand notation—using just the root name. For example, the math library is `libm.so` or `libm.a`. To link it in simply type `gcc filename.c -lm`. The only difference between typing this and explicitly using the library name as an input file (e.g., `gcc -o runfile filename.c libm.a`) is the use of shorthand notation with the `-l` option. Furthermore, the placement of the library in the list of input files makes a difference—input files are scanned for linking in the order they are listed. The notation above says "scan `filename.o` first, then `libm.a`" [52].

In MS-Windows environments, shared libraries are called *dynamic link libraries* and have the suffix `.dll` rather than `.so`. Similarly, static libraries have the suffix `.lib` rather than `.a`. Because of this, the shorthand notation `gcc` uses in linking libraries is useful only if you follow the UNIX library name standard (GNU's ported standard libraries do).

The above constitutes only a small portion of the multitude of options available in `gcc`. However, these are the ones you'll find yourself using all the time. If you need a special option not defined here, see the "man" or "info" page for `gcc`, or consult [52].

B.2 Using `make` and `Makefiles`

GNU's `gcc` compiler is quite a powerful development tool, but it certainly does not have answers for everything. Writing short programs with just a couple of source code files and a couple of header files, `gcc` may be the only programming tool you need. However, if you ever need to deal with large programs, which consist of many input files that exist in many different directories, it becomes much more difficult to keep files organized. Imagine trying to create a single executable from, say, even 10 source code files. You could use the `-c` option of `gcc` and create a single object file for each of the source code files (i.e., create 10 object files). Then these can be linked together in an additional `gcc` call with all 10 object files as arguments to `gcc`. That way, when you change only one of the source code files, you only need to compile that one file and then just relink all the object files to create the updated executable file.

You probably see that this becomes quite tedious. Things get more complicated when each of the source code files has its own header file. Now, we have added 10 additional files. If this isn't enough, we have only considered so far where each source code depends only on itself. This is not realistic, as often separate source code files are dependent on other header files for the definitions of constants, classes, and structures.[1] They are often dependent on other source code files as well for certain functions. So now we have 20 total files, each of which is interdependent on each other. Now if one file is changed, you may have to recompile two, three, four, or more files so that the changes are passed completely through to the executable. The `make` utility *automatically* determines which files are old and runs the commands necessary to update them. This frees the developer from determining which files need to be recompiled due to changes made in the code (not to mention from compiling as well). This makes development of code less error prone by eliminating the possibility of the developer forgetting to recompile a necessary file. Simply by typing `make` you can have the entire project compiled and updated, if necessary.

Actually, most commercial C and C++ development tools use some sort of make utility.[2] Here, we describe GNU's version of make, which like `gcc` is quite a powerful program. It offers many added functions over other commercial versions. This section describes just some introductory information about `make` and writing makefiles. There is a document of over 200 pages that describes GNU's `make`; interested readers can consult this for a more detailed description of the advanced features [51].

---

[1]The term *dependencies* is used for the files that a particular program depends on.

[2]You can write makefiles for both Borland C and MS Visual C.

B.2.1 How to Write Makefiles

In order to use `make`, it is necessary to write a `makefile` that essentially tells `make` what to do. Makefiles contain five things: explicit rules, implicit rules, directives, and comments. A *rule* tells `make` what to do, how to do it, and when to do it. It follows the syntax

```
target ... : dependencies ...
        commands
        ...
```

The `target` is usually the name of the file that is created due to the `commands` that are executed. The `dependencies` are, of course, the files that the target depends on. By looking at the time and date when `target` was created, and the dates and times when the dependencies were created, `make` can determine whether or not `target` needs to be rebuilt or remade. If one of the dependencies is newer than the target, it is remade. Similarly, if `target` does not exist, it is remade. When a target is remade, `make` simply goes through the `commands` one line at a time until the end of the rule is reached. Note that a TAB character must precede each command.

The easiest way to see how to write one of these is to look at an example. A makefile written for an experiment in our laboratory (i.e., a rotational inverted pendulum) is shown in Example B.1.

Example B.1: Makefile for the pendulum

```
##########################################################################
# This Makefile is for the Pendulum Experiment.
##########################################################################
# For Linux:
# To run, in the directory the Makefile is located type 'make'.  Type
# 'make clean' to delete the object files.
#
# Requirements --- Uses GNU make and gcc
##########################################################################

all: pendulum

pendulum: pendulum.o param.o hctl.o
        gcc -o pendulum pendulum.o param.o hctl.o -ld20

pendulum.o: pendulum.c pendulum.h param.h hctl.h das20.h
        gcc -Wall -O2 -c -o pendulum.o pendulum.c

param.o: param.c param.h pendulum.h
        gcc -Wall -O2 -c -o param.o param.c

hctl.o: hctl.c hctl.h das20.h
        gcc -Wall -O2 -c -o hctl.o hctl.c

clean:
        -rm -f pendulum.o param.o hctl.o das20.o pendulum
```

Looking at this file, you can see that it is made up of five targets. The comment lines all begin with the character #. By default, make looks for the makefile in the current directory by trying the following names, in order—GNUmakefile, makefile, Makefile. Normally, you should call your file Makefile unless it is GNU-specific, in which case it may be advantageous to use GNUmakefile. Once the makefile is found, make starts with the first target, in this case all, and determines whether or not it should be remade by looking at its dependencies. In this case, all has only one dependency, the executable file pendulum. So make looks then at the dependencies of pendulum to determine if it needs to be re-made. Now we find that pendulum depends on three object files. make proceeds to go through each of the object files to determine if they need to be re-made first (by looking at the dependencies of each object file). If none of the object files need to be updated, and pendulum is at least as "new" as the newest version of the object files, then pendulum is up-to-date and need not be changed. Thus, make returns, letting the caller know that everything is up to date with the friendly message

```
make:  Nothing to be done for 'all'
```

Note that if the target filename does not exist, then it is automatically remade, since make attempts to determine if the target name is as new as its dependencies and cannot find the file associated with the target.

On the other hand, say everything was up to date, then we edit the file pendulum.h. Upon rerunning, make would find that the file pendulum.o was out-of-date when compared to pendulum.h and executes the commands contained in that rule. In this case, make would run

```
gcc -Wall -O2 -c -o pendulum.o pendulum.c
```

producing an updated object file. The same would happen for param.o. The file hctl.o would *not* be changed, however, since it does not depend on pendulum.h; make sees that it is up-to-date with respect to its dependencies and skips the commands for this target. Then it would return to the pendulum target and find that the two object files are now newer than the target, and thus execute the command

```
gcc -o pendulum pendulum.o param.o hctl.o -ld20
```

producing the updated executable automatically.

The target clean is a special target in that it does not have any dependencies and its commands do not produce a file named "clean." It is what we will call a *phony* target. Because there are no dependencies, and it does not create a filename called "clean," it is always executed when you type

```
make clean
```

Note, however, that it will *not* execute when we type make or make all since none of the other files depend on clean. By examining the commands of clean, it is easy to see what it does—removes (or "cleans") the directory of the object and executable files. Sometimes it is nice to start over from a clean slate, and as a result you will see this command often in makefiles. In this case, if you type "make clean; make," then all of the files will be remade since you first removed them. The minus sign in front of -rm -f ... tells make to ignore any errors that may be produced when invoking that command (such as if a file did not exist). Note from this example that you can specify the name of the target that you want to make. Instead of typing make above, we could also type make pendulum and the same thing will happen. The target all is often placed in makefiles, since many times make produces more than one executable. In this case it is easy to place several executables as the dependencies for all, and type make all to recompile everything.

B.2.2 Using Variables to Make Makefiles Easier to Read

In this section we discuss the use of variables that will result in makefiles that are easier to read.

User-Defined Variables

Within makefiles you can specify variables that can be used within the makefile. This is simply done in the following fashion:

```
SRCDIR = ./src
```

This assigns the value of "./src" to the variable SRCDIR. Anytime you wish to use this variable, type $(SRCDIR) within the makefile and make will substitute the value "./src" at each occurrence. Note that because make uses the dollar sign in this fashion, to get the effect of a dollar sign in a command (such as in a shell script) you must type "$$."

The above illustrates one useful example of using variables. Often, in large programs, the source code files are located in their own directory, as are include files and object files. By using variables instead of explicitly typing the directory structure each time, makefiles become easier to write and much easier to change. If you decide later to change the location of all the source files, then you only need to change one value in the makefile.

Another popular example of the use of variables is the assigning of compiler and compiler flags. Note that in the previous makefile example, we use the same common flags when invoking gcc. We could assign a variable to this as well. For example, we could have done

```
CC = gcc
CFLAGS = -Wall -O2
```

and used `$(CC) $(CFLAGS) -c -o pendulum.o pendulum.c` when writing the rule for `pendulum.o`.[3]

Automatic Variables

`make` provides special variables it assigns for you to use at your discretion. These are called *automatic variables.* You use these just as any other variable; you need not worry about confusing them; most of these variables are assigned to special characters you do not normally use. Here is a list of the common ones:

`<` This variable refers to the first named file in the dependency list. If a rule were in the form

```
pendulum.o: pendulum.c pendulum.h param.h hctl.h das20.h
```

then the value of the variable `$<` used *within that rule* would be the file `pendulum.c`. Note that this is local to each rule (and of course holds a different value in each rule).

`^` This is similar to `<` except that it refers to *all* of the dependencies of the particular target. This is also local to each rule. In the preceding example, the value of `$^` would be `pendulum.c pendulum.h param.h hctl.h das20.h`.

`@` This variable is also local to each rule and is assigned the target name. Using this in the example rule above, `$@` would be assigned `pendulum.o`

`*` This is assigned the *root* of the target filename, that is, the name of the target without the extension (in GNU `make`, also without the directory structure). In the example above, `$*` is assigned the value `pendulum`.

There are a few other automatic variables, but they are used less frequently.

Example of the Use of Variables

To show an example of the use of both automatic and user-defined variables, included here is the previous rotational inverted pendulum makefile rewritten to incorporate variables. First, we move the source files into each of their own directories to keep things organized—source files are moved into `./src`, headers into `./include`, objects into `./lib`, and executables in `./bin`. The updated makefile is shown in Example B.2.

[3]If you decide to delve more deeply into `make`, you will find that it actually uses these variables in its *implicit rules*. `make` is relatively smart—it already knows how to produce, say, a `.o` file from a `.c` or `.cc` file using built in implicit rules. The default is for `make` to use "cc -c" for C programs and "c++ -c" for C++ programs. You can change these defaults using the variables `CC` and `CFLAGS` (for the C language) or `CPP` and `CPPFLAGS` (for C++).

Example B.2: Better makefile for the pendulum

```
###########################################################################
# This Makefile is for the Pendulum Experiment.
###########################################################################
# For Linux:
# To run, in the directory the Makefile is located type 'make'.  Type
# 'make clean' to delete the object files.
#
# Requirements --- Uses GNU make and gcc
###########################################################################

# Directories for file locations. THESE MAY HAVE TO BE CHANGED if you
# change the directory structure
TOPDIR := $(shell if [ "$$PWD" != "" ]; then echo $$PWD; else pwd; fi)
SRCDIR = src
INCDIR = include
BINDIR = bin
LIBDIR = lib

# Compiler and Compiler flags
CC = gcc
CFLAGS = -O2 -I$(INCDIR) -Wall

# Files
SRCS = \
        $(SRCDIR)/hctl.c \
        $(SRCDIR)/param.c \
        $(SRCDIR)/pendulum.c
HEADERS = \
        $(INCDIR)/das20.h \
        $(INCDIR)/hctl.h \
        $(INCDIR)/param.h \
        $(INCDIR)/pendulum.h
OBJS = \
        $(LIBDIR)/hctl.o \
        $(LIBDIR)/param.o \
        $(LIBDIR)/pendulum.o
BINS = \
        $(BINDIR)/pendulum

# Default target
all: $(BINS)

# Other targets/rules
$(BINDIR)/pendulum: $(OBJS)
        $(CC) -o$@ $^ -ld20

$(LIBDIR)/hctl.o: $(SRCDIR)/hctl.c $(INCDIR)/hctl.h $(INCDIR)/das20.h
        $(CC) $(CFLAGS) -c -o$@ $<

$(LIBDIR)/param.o: $(SRCDIR)/param.c $(INCDIR)/param.h $(INCDIR)/pendulum.h
        $(CC) $(CFLAGS) -c -o$@ $<

$(LIBDIR)/pendulum.o: $(SRCDIR)/pendulum.c $(INCDIR)/das20.h $(INCDIR)/pendulum.h \
        $(INCDIR)/param.h $(INCDIR)/hctl.h
        $(CC) $(CFLAGS) -c -o$@ $<
```

```
.PHONY: clean

clean:
        -rm -f $(OBJS) $(BINS)
```

Although it is somewhat longer, it provides a simpler interface to work with. The variables SRCS and HEADERS are not used explicitly, but are included so that, at a quick glance, you can tell what files the project uses. The backslashes included in the lines tell make to treat the series of lines as one long line of files. (*Note:* You cannot just simply hit enter in the middle of a variable definition or dependency list, because make assumes this is the end of that particular input. You must use a backslash at the end of each line if that definition/dependency list continues to the next line.)

Using variables is a convenient way to make your makefiles more readable. As a project becomes more involved (and contains a larger number of files), the use of automatic variables makes things significantly easier.

B.3 RCS Makefiles

In the short introduction above, we just touched upon the options and abilities of GNU's two software tools—gcc and make. In this section we introduce the RCS makefiles which will ease the job of the RCS application developer in compilation and linking of the code. Readers interested in learning more about GNU's gcc and make may consult [51, 52]. When designing RCS applications in UNIX-type environments, these are important tools. This is because, in general, RCS applications are developed by several users on several different computers or platforms and require compilation and linking of several different files, and therefore it is very convenient to use makefiles for this purpose. Each specific application and particular platform needs its own set of rules and therefore its own makefile. However, the programmer should not be worried about them because it is possible to generate the makefiles needed by the application using the RCS design tool.

B.3.1 Makefile *generic.def*

The file *generic.def* (formerly called *Makefile.generic*) is a general RCS makefile which provides definitions that are not application specific. It provides the following important features:

1. Allows the application to be easily compiled for multiple platforms.

2. Provides each programmer with the ability to develop and test modifications to the application without affecting other programmers working on the application or users.

3. Allows information about the compilers and libraries available at a given site for each platform to be centralized.

4. Allows programmers more easily to generate and use automatically generated dependency rules for header files.

5. Adds the directory for the RCS library header files to the compiler's include path.

In order for *generic.def* to work, applications need to follow the particular directory structure which was discussed in Chapter 10 (refer to Table 10.1) and can be generated automatically by the RCS design tool. In addition to that, the programmer needs to set some makefile variables and follow the RCS design tool conventions for filenames. The file *generic.def* is compatible with GNU's `make` but is not compatible with some other make utilities such as Microsoft's `NMAKE`.

B.3.2 Application Include Makefile

It is recommended that each application create a short makefile to include some definitions to be used by makefiles throughout the application. This file should be a very convenient place to include *generic.def* and to define the `APPDIR` variable which *generic.def* will use. `APPDIR` should be set to the full path to the main application directory. This may also be a good place to set `LOCAL_CFLAGS` or other variables used by *generic.def*. An example file that will be called *Makefile.inc* in later examples is shown in Example B.3.

Example B.3: Makefile for the intelligent vehicle application

```
# This makefile should define the APPDIR any other
# variables used thoughout the application such as
# LOCAL_CFLAGS, or RCSLIB_MAIN_DIR and then include the
# RCS generic makefile definitions.

# Set the Release Directory for this Application
APPDIR = /proj/ivhs

# Set the Development Directory for this user
USER_DIR = /home/veys/ivhs

# Set the main RCS Library Directory
RCSLIB_MAIN_DIR = /usr/local/rcslib

# Include the generic RCS makefile definitions
include $(RCSLIB_MAIN_DIR)/etc/generic.def
```

This file should be placed in the top-level directory in order to follow the RCS design tool convention.

B.3.3 Top-Level Makefile

The primary purpose of the top-level makefile is to provide a convenient way to call for a make of the appropriate target(s) in each subdirectory. There are several targets of importance in *generic.def* that can easily be passed to the subdirectories, as Example B.4 shows.

Example B.4: Top-level makefile for the intelligent vehicle

```
# Top-Level Makefile
# This makefile should call for makes in each of the
# src subdirctories in this application

# Include Application Specific Definions and the RCS generic Makefile
include Makefile.inc

# Rules to go down into subdirectories
all clean depend sources headers:
        ( cd src/intf; $(MAKE) $@;)
        ( cd src/util; $(MAKE) $@;)
        ( cd src/attention; $(MAKE) $@;)
        ( cd src/drive; $(MAKE) $@;)
        ( cd src/maneuver; $(MAKE) $@;)
        ( cd src/supervisor; $(MAKE) $@;)
        ( cd src/main; $(MAKE) $@;)
```

Note from this file that all the rules are specified in the same manner. They do not have any dependencies specified, but call the makefiles in the subdirectories specified. Therefore, typing `make all` on the command line is equivalent to typing `make all` in each of the `src/intf`, `src/util`, `src/attention`, `src/drive`, `src/maneuver`, `src/supervisor`, and `src/main` subdirectories; similarly for the other targets, `clean`, `depend`, `sources`, and `headers`.

B.3.4 Subdirectory Makefiles

Each source subdirectory should have its own Makefile. Within this makefile programmers will need to define several variables, include the application include makefile, and provide rules for linking any binaries or libraries together. Example B.5 shows such a file.

Example B.5: Subdirectory makefile for the drive module of an intelligent vehicle

```
# This makefile should define the sources, headers and objects
# stored or created with this directory and then include
# the application include Makefile.

SRCS =  \
        drive.cc \

HEADERS = \
```

```
        drive.hh \

OBJS = \
        drive.o \

LIBS =

BINS =

# Include Application Specific Definitions and the RCS generic Makefile
include ../../Makefile.inc
```

Note that the file includes the application makefile with the application-specific definitions *Makefile.inc* and therefore also *generic.def*, where the rules for creating the object files from the sources and headers defined within this makefile are defined. Since the files here are the code for the module itself and not the code for the main program, we do not see any rule for binary executables. In Example B.6 we provide a second example of a subdirectory file which creates the main executables needed by an intelligent vehicle (discussed throughout this book).

Example B.6: Subdirectory makefile for the intelligent vehicle

```
# This makefile should define the sources, headers and objects
# stored or created with this directory and then include
# the application include Makefile.

SRCS =  \
        ivhsmain.cc \
        ivhssvr.cc \
        servomain.cc \

HEADERS = \

OBJS = \
        ivhsmain.o \
        ivhssvr.o \
        servomain.o \

ifeq (vxworks,$(findstring vxworks,$(PLAT)))

LIBS =

else
LIBS = libivhs.a

endif

ifeq (vxworks,$(findstring vxworks,$(PLAT)))

BINS =
```

```
else

BINS = \
        ivhssvr \
        ivhsmain \
        servomain \

endif

# Include Application Specific Definitions and the RCS generic Makefile
include ../../Makefile.inc

# Rules for specific libraries

# ivhs.a
$(DEVP_LIB_DIR)/libivhs.a: \
        $(DEVP_LIB_DIR)/veh_sim.o \
        $(DEVP_LIB_DIR)/rk4_sim.o \
        $(DEVP_LIB_DIR)/attn_to_supn.o \
        $(DEVP_LIB_DIR)/attn_to_drvn.o \
        $(DEVP_LIB_DIR)/attention.o \
        $(DEVP_LIB_DIR)/drive.o \
        $(DEVP_LIB_DIR)/maneuver.o \
        $(DEVP_LIB_DIR)/supervisor.o \
        $(DEVP_LIB_DIR)/attentionn.o \
        $(DEVP_LIB_DIR)/driven.o \
        $(DEVP_LIB_DIR)/maneuvern.o \
        $(DEVP_LIB_DIR)/supervisorn.o
        $(COMPILER_SETUP); \
        $(AR) cr $@ $^
        $(RANLIB) $@

# Rules for specific binaries

# ivhssvr
$(DEVP_BIN_DIR)/ivhssvr: $(DEVP_LIB_DIR)/ivhssvr.o \
        $(DEVP_LIB_DIR)/libivhs.a
        $(COMPILER_SETUP); \
        $(CPLUSPLUS) $(CPLUSPLUSFLAGS) $^ \
        $(CPLUSPLUSLINK) $(EXTRA_LIBS) \
        $(RCS_LIBRARY) \
        -o $@

ivhssvr: $(DEVP_BIN_DIR)/ivhssvr

.PHONY: ivhssvr

# ivhsmain
$(DEVP_BIN_DIR)/ivhsmain: $(DEVP_LIB_DIR)/ivhsmain.o \
        $(DEVP_LIB_DIR)/libivhs.a
        $(COMPILER_SETUP); \
        $(CPLUSPLUS) $(CPLUSPLUSFLAGS) $^ \
        $(CPLUSPLUSLINK) $(EXTRA_LIBS) \
        $(RCS_LIBRARY) \
        -o $@

ivhsmain: $(DEVP_BIN_DIR)/ivhsmain
```

```
.PHONY: ivhsmain

# servomain
$(DEVP_BIN_DIR)/servomain: $(DEVP_LIB_DIR)/servomain.o \
        $(DEVP_LIB_DIR)/libivhs.a
        $(COMPILER_SETUP); \
        $(CPLUSPLUS) $(CPLUSPLUSFLAGS) $^ \
        $(CPLUSPLUSLINK) $(EXTRA_LIBS) \
        $(RCS_LIBRARY) \
        -o $@

servomain: $(DEVP_BIN_DIR)/servomain

.PHONY: servomain
```

Note that this makefile first creates a library called `libivhs.a` using the already existing object files and then compiles and links all the executables needed by an intelligent vehicle, including the NML server `ivhssvr` needed for communications.

The variables in the files above such as `DEVP_LIB_DIR` and `COMPILER_SETUP`, are defined in the *generic.def* or in the platform-specific makefile. We discuss those of them which are most needed by the application programmer in the next section.

B.3.5 Variables Used in *generic.def*

Several variables are used within *generic.def* to control how your application is built. Other variables are set by *generic.def* or within one of the platform-specific makefiles of the form *$(PLAT).def*, such as *vxworksCC.def* or *sunos4.def* (formerly called *Makefile.vxworksCC* and *Makefile.sunos4*) to aid in the development of rules for specific binaries and libraries. Below, we provide a list of the variables used in *generic.def*, together with the meaning of this variable and the recomended place to set this variable. If the place for setting is *$(PLAT).def* or *generic.def*, the variable must be set there and should not be set anywhere else. The platform-specific makefiles are included automatically by *generic.def*, so there is no need to include them directly from application makefiles.

- `APPDIR` – Set in application include makefile
  Specifies the main application directory, which is used in the `INCLUDE` path and for performing an install when the application is ready to be released.

- `AR` – Set in *$(PLAT).def*
  Provides the name and path of the library archiver for the given platform. (For more information, see the `ar` "man" page.)

- `BINS` – Set in subdirectory makefile
  Lists the executable binary files that should be created by running make

in this subdirectory. Binaries will need to have a separate rule at the end of the makefile.

- `CC` – Set in *$(PLAT).def*
  Provides the name and path of the C compiler for the given platform.

- `CPLUSPLUS` – Set in *$(PLAT).def*
  Provides the name and path of the C++ compiler for the given platform.

- `COMPILER_SETUP` – Set in *$(PLAT).def*
  Provides the command(s) required to set up the environment so that the compiler and related tools can be used. It should be used at the beginning of commands in rules for specific binaries and libraries.

- `DEVP_BIN_DIR` – Set in *generic.def*
  Provides the name of the directory where executable binary files should land within the programmer's workspace. It should be used to provide rules for specific binaries.

- `DEVP_LIB_DIR` – Set in *generic.def*
  Provides the name of the directory where object files and libraries should land within the programmer's workspace. It should be used to provide rules for specific binaries and libraries.

- `HEADERS` – Set in subdirectory makefile
  Lists the header files found in this directory that should be copied to the platform include directory so that they may be used from other directories or by other programmers.

- `LIBS` – Set in subdirectory makefile
  Lists the library files that should be created by running make in this subdirectory. Libraries will need to have a separate rule at the end of the makefile.

- `LOCAL_CFLAGS` – Set in application include makefile
  Lists options that will be passed to the C or C++ compiler.

- `LOCAL_CPLUSPLUSFLAGS` – Set in application include makefile
  Lists options that will be passed to the C++ compiler, but not to a C compiler.

- `OBJS` – Set in subdirectory makefile
  Lists the object files that should be compiled by running make in this subdirectory.

- `PLAT` – Set in command line
  Specifies which platform to compile for. It is normally set on the command line so that programmers can easily switch between multiple compilers and cross-compilers. It defaults to the value of the `osrev` environment variable.

- `RANLIB` – Set in *$(PLAT).def*
  Provides the name and path of the `ranlib` utility for the given platform if one exists and is required; otherwise, the variable contains some command that should have no effect. The `ranlib` utility converts archives to random libraries. (For more info, see the `ranlib` "man" page.)

- `RCS_INCLUDE` – Set in *generic.def*
  Provides the directory where header files for the RCS Library are placed.

- `RCS_LIBRARY` – Set in *generic.def*
  Provides the name and path of the RCS library for the platform set with `PLAT`.

- `RCS_PLATLIB` – Set in *generic.def*
  Provides the path to the RCS library for the platform set with `PLAT`. This is also the location of the `pmac` and `pcio` libraries if they exist for the platform.

- `SOURCES` - Set in subdirectory makefile
  Lists the C and C++ files used from that subdirectory. C files should have the `.c` extension and C++ files should have the `.cc` extension. The list is used for creating automatic dependency lists and for storing archives of the source code during an install. Do not list header files here.

- `USER_DIR` - Set in programmer's environment, application include makefile, or command line
  Specifies the top-level directory in the programmer's workspace. It defaults to the name of the application under the programmer's home directory. It is recommended that the applications be set up so that the default can be used and `USER_DIR` need not be set. For example, if `APPDIR=/home/manta/emc` and the programmer's login name was "shackle," then the workspace should be placed `~shackle/emc` to match the default.

B.3.6 Special Targets

The main makefile *generic.def* provides rules for creating several `PHONY` targets. By specifying one of these targets on the command line, the programmer can have several useful tasks performed. Next, we list these for the reader.

- `all`
  Update everything within the programmer's workspace, including copying header files to the `include` directory, compiling, linking, and archiving as necessary.

- `clean`
  Deletes object files and programs from the programmer's workspace. This is useful, for example, when a change in compiler options would not take effect unless the entire application is rebuilt.

- `depend`
  Generates lists of dependencies so that files will be recompiled whenever a header file they include directly or indirectly has changed (uses the `makedepend` utility).

- `headers`
  Copies the header files to the `src` and `include` subdirectories of the current platform directory. For Windows and DOS platforms the files are also converted with `unix2dos`.

- `install`
  Copies all the source, header, object files, libraries, and programs from the programmer's workspace to the main application release directory.

- `sources`
  Copies the source and header files to the `src` and `include` subdirectories of the current platform directory. For Windows and DOS platforms the files are also converted with `unix2dos`, and the `.cc` extension is replaced with `.cpp`.

This concludes our discussion of GNU's make and RCS makefiles. The programmer is not required to follow the particular directory structure generated by the RCS design tool or the RCS makefiles. However, we recommend that you do, because this often makes the program development easier.

Appendix C

General Operating System Concepts

C.1 Computer Network Communication Protocols

Communication networks have a layered structure. In the Open System Interconnection (OSI) reference model proposed by the International Standards Organization (ISO), there are seven layers in a network: the *physical layer, data link layer, network layer, transport layer, session layer, presentation layer*, and *application layer*. Every layer provides its own functions, independent of the others. In other words, changing protocols or techniques of one layer does not affect the functions of other layers. Each layer also has a set of protocols (rules) to communicate with its neighboring layers and with the same layer on a different machine. *Transmission Control Protocol* (TCP) and *User Datagram Protocol* (UDP) are two protocols of the *transport layer*. Transport layer protocols involve cooperation only between two nodes at each end of a communication path rather than cooperation among all nodes in the path. They also put more emphasis on naming and addressing than do protocols in other layers. TCP and UDP are generally used with *Internet Protocol (IP)*, which is a *network layer* protocol.

C.1.1 TCP

TCP is a connection-oriented protocol that maintains end-to-end reliability over a full duplex connection (i.e., both communicating machines send and receive information). It provides reliable service that guarantees delivery of a stream of data from one machine to another without duplication or data loss despite unreliable packet delivery service of the *network layer*. It is usually implemented with IP; however, since it assumes little about the underlying communication system, it can be used with a variety of packet delivery systems. For instance, it can be implemented to use dial-up telephone lines, a local area network, a high-speed fiber optic network, or a lower-speed long network.

TCP *data units* consist of *header* and *data* parts. The minimum header length is 20 bytes; it is composed of *local port number, remote port number, sequence number, acknowledgment number, length, checksum,* and other fields related to establishing/closing connection, priority, and so on. Port numbers should be unique so that the data of different processes are not confused. "Sequence number" is the number of the current packet, "acknowledgment number" is the number of the last packet received from the remote port plus the number of bytes in that packet, "length" indicates the number of bytes in the data unit (including the header), and "checksum" is used for detecting errors and determining if the packet has arrived in the right port.

In TCP, sent packets which are not acknowledged by the remote port in a predefined period of time (timeout) are retransmitted. Typical TCP communication process can be divided into three phases:

- Establishing connection
- Transmitting and receiving data
- Closing connection

C.1.2 UDP

UDP is a connectionless *transport layer* protocol. UDP *data units,* called *datagrams,* also consist of *header* and *data* fields. The header parts of datagrams are smaller than those of TCP data units. They consist of *source port number, destination port number, length,* and *checksum* fields. As in TCP, "port numbers" are used to distinguish the given communicating process among other processes in the machine. "Length" gives the number of bytes in the datagram, and "checksum" is for detecting possible errors and the correct ports. The difference from TCP is that there is no error reporting in UDP. Moreover, there is no connection establishment between communicating hosts. Therefore, received datagrams are not acknowledged to the sending host. For this reason, lost or corrupted data is not resent by the sending host, and the receiving end does not know if there are lost packets and/or corrupted data. UDP simply forms a datagram and transfers it to the network layer (IP) for transmission. The receiving end uses the checksum to see if incoming datagrams belong to that port and if there is an error, passes the valid ones to the specified port of the upper layer and simply discards the invalid ones.

If we compare TCP and UDP, we see that TCP is more reliable than UDP, because of its error correction (retransmitting) capability. However, this capability has also a drawback: It causes TCP to be much slower than UDP. Moreover, its packets have large headers. Even 1 byte of data needs 20 bytes of header in TCP, which is a big overhead that lowers the efficiency of the network.

TCP and UDP are two important protocols which are used in the RCS library for communication of processes with remote buffers (i.e., buffers which are located on remote hosts or computers).

C.2 Operating System Concepts

An operating system is a program which provides an interface between the user and the computer hardware. It provides an orderly and controlled allocation of the processors, memories, and input/output devices among various user programs competing for them. When the computer has more than one user, the need for managing and protecting memory, input/output devices, and other resources is greater. The interface between user programs and the operating system is provided by a set of instructions, called system calls, which may vary from system to system.

C.2.1 Processes (Tasks) and Multitasking

In the literature on operating systems, *task* or *process* is defined as a program in execution. Every process has its address space, which is a part of the memory, from which the process can write and read. This memory space is divided into *program space* and *data space*. Program space contains the executable itself and the data space is the memory where the variables (data) of the program are defined. Processes using the same device (resource) in the same computer and/or reading and writing to the same memory space are referred to as *concurrent processes*. It is a job of the operating system to allocate the shared resources to the requesting programs in the order given. For example, consider two tasks that are using the same buffer, such that one of them reads the buffer and the other one writes to the buffer. The operating system should prevent the writer process from writing the buffer while the reader is reading, or vice versa; otherwise, the reader may read wrong data. Similarly, if two processes request a hardware device (e.g., a printer) simultaneously, they should be granted with access to it sequentially.

The concept of running more than one process on one processor simultaneously is called *multitasking*. The part of the operating system responsible for switching the processor (allocating the processor) from one process to another is referred to as a *process scheduler*. It loads a process to the processor, allows it to run for a while, and then takes it from the processor (preempts it) and loads another process. This is done in the background, and the user perceives the system as if it were running all the tasks simultaneously.

C.2.2 Critical Sections and Mutual Exclusion

Consider again the the reader and writer processes example. Assume that the writer began to write into the buffer (it wrote some data but didn't finish), but it was preempted by the scheduler and the reader was loaded. If the reader process accesses the buffer, it will read erroneous data. Such conditions are referred as *race conditions* and the sections in the programs in which the program accesses a shared resource (memory, device, etc.) are called *critical sections* or *critical regions*. In order to avoid race conditions, processes should be prevented from entering their critical region simultaneously. This concept is referred to as *mutual exclusion*.

There are several techniques for mutual exclusion of tasks. One way for mutual exclusion is using a *semaphore*. The main idea in this method is to have a semaphore variable and two indivisible operations (i.e., the operation cannot be divided; therefore, the process scheduler is not allowed to preempt a process performing this operation before completing it) for incrementing and decrementing the value of the semaphore. The decrementation operation is done as long as the value of the semaphore is greater than zero; otherwise, the process waits (or it is sent to sleep) until the value of the semaphore becomes greater than zero. Consider a binary semaphore (with values of either 0 or 1). Every process decrements the value of the semaphore (takes the semaphore) before entering its critical section and increments it (releases it) on exit from the critical region. In that case if there is a process already in its critical region, the other process can't enter its critical region, because the semaphore is taken by the first process. The new task has to wait until the semaphore is released.

Another method for mutual exclusion of concurrent processes is *disabling–enabling interrupts*. When a critical section is entered, the interrupts are disabled, which prevents the process scheduler from switching the current task. Therefore, there is no other process running on the processor and there cannot be a conflict. Interrupts are enabled on exit from the critical section. This approach has some drawbacks, because it does not work in multiprocessor systems with shared memory, since it disables interrupts of one of the processors only, whereas the others will still be able to access the shared device or memory. Moreover, since the user programs have access to the interrupts, this approach could be dangerous. For this reason, instead of using this approach, its idea is used in implementing semaphores.

Another way for implementing semaphores is to use a *test and set* instruction (if supported by the hardware of the system), which performs testing of a variable and setting it to a given value in a single indivisible instruction.

Semaphores are usually provided by the operating system as operating system calls or as built-in functions and types in a system implementation language.

Appendix D

RCS Version Functions

The RCS library is still under development. Therefore, during the development progress, different versions of the library appear. These are described by numbers called *version numbers.* The RCS version numbers are two decimal numbers separated by a decimal point. The first number is called the *major version number* and the second one is called the *minor version number.* For instance, in 2.7 the major version is 2 and the minor version is 7. The major number reflects major modification in the library and the minor number is for less important changes. Therefore, there should not be a problem in the communication of two programs which differ only in minor version number. However, if they differ in major version number, then you may need to get the new version for one of them. The macro that defines the version of a given library is `RCS_VERSION`. It is defined in `rcs.hh` and is in the form $m.n$, where m is the major version number and n is the minor version number.

There are several ways for testing the version of a given library. The function `print_rcs_version` prints a line describing the RCS library, its version, date compiled, and so on. There is a function for comparing the version of the RCS library linked to a particular version. This function is called `rcs_version_compare` and its declaration is as follows:

```
int rcs_version_compare(const char * compversion);
```

It returns 0 if the version string sent to it (i.e., `compversion`) matches the version of the library, -1 if the RCS library linked in is older than `compversion`, and +1 if the RCS library linked in is newer than `compversion`. If no minor number is included in `compversion`, any version with the given major number will return 0. Example D.1 shows how this function can be used.

Example D.1: Using the function `rcs_version_compare` for finding the version of an RCS library

```
main()
{
        if(rcs_version_compare("2.5") < 0)
```

```
        {
                printf("RCS library older than 2.5\n");
        }
}

// Only compare major number.
main()
{
        if(rcs_version_compare("2") != 0)
        {
                printf("RCS library is not version 2.\n");
        }
        else
        {
                printf("RCS library is version 2.
                        (It could be 2.0 or 2.99, we don't care.)\n");
        }
}
```

If you suspect that the version of the RCS library linked and that of the `rcs.hh` included in the application do not match, you can test this as shown in Example D.2.

Example D.2: Testing for version mismatch between `rcs.hh` and the RCS library included

```
// This example shows how to test if the rcs.hh included matches the
// library that is linked in.
main()
{
        if(rcs_version_compare(RCS_VERSION) != 0)
        {
                printf("RCS library and header file don't match.\n");
        }
}
```

Another way to determine the version of a library linked into an executable is to use the SCCS `what` command. This is shown in Example D.3.

Example D.3: Using system command to find the version number of a library linked into an executable

The command

```
what executable_file | grep RCS_LIBRARY_VERSION
```

will give an output similar to

```
RCS_LIBRARY_VERSION: 2.9 Compiled on  May  6 1997 at 15:55:35 for the
sunos5 platform.
```

Appendix E

Platforms Tested

Machine Type	Operating System	Compiler	Additional Software	Platform Name
486/Pentium	MS-DOS	Borland C++	PC-NFS	`dos_bor`
486/Pentium	MS-DOS	Microsoft C++	PC-NFS	`dos_msc`
486/Pentium	16-bit Windows	Borland C++	PC-NFS	`win16bor`
486/Pentium	16-bit Windows	Microsoft C++	PC-NFS	`win16msc`
486/Pentium	32-bit Windows	Borland C++	PC-NFS	`win32bor`
486/Pentium	32-bit Windows	Microsoft C++	PC-NFS	`win32msc`
486/Pentium	LynxOs 2.1-2.2	GNU g++	included with OS	`lynxosPC`
486/Pentium	LynxOs 2.3	GNU g++	included with OS	`lynxosPC23`
486/Pentium	Linux 2.0.3	GNU g++	included with OS	`linux`
MVME 167	LynxOs	GNU g++	included with OS	`lynxosVME`
Heurikon HK68/V3D	VxWorks 5.1	GNU g++	included with OS	`vxworks5.1`
MVME 167,162,147	VxWorks 5.1	GNU g++	included with OS	`vxworks5.1`
MVME 162	VxWorks 5.3	GNU g++	included with OS	`vxworks5.3`

Machine Type	Operating System	Compiler	Additional Software	Platform Name
CETIA (PowerPC)	VxWorks 5.3	GNU `g++`	included with OS	`vxworksppc5.3`
Sun Work-Station	SunOs 4 or 5	GNU `g++`	included with OS	`sunos4` or `sunos5`
Sun Work-Station	SunOs 5	CenterLine `CC`	included with OS	`sunos5CC`
Sun Work-Station	SunOs 5	Sparcworks	included with OS	`sparcworks_sun5`
SGI Work-Station	IRIX 5.x	GNU `g++`	included with OS	`irix5`
SGI Work-Station	IRIX 6.x	GNU `g++`	included with OS	`irix6`

Bibliography

[1] J. Ackermann, J. Guldner, W. Sienel, R. Steinhauser, and V. Utkin. Linear and nonlinear control design for robust automatic steering. *IEEE Trans. on Control Systems Technology*, 3(1):132–143, March 1995.

[2] J. S. Albus. System description and design architecture for multiple undersea vehicles. *Technical Note 1251*. National Institute of Standards and Technology, Gaithersburg, MD, September 1988.

[3] J. S. Albus, E. Barkmeyer, and A. Jones. Approach to a system architecture for post office automation. In *Proc. of the USPS 4th Advanced Technology Conference*, Washington, DC, November 5–7 1991.

[4] J. S. Albus, M. Juberts, and S. Szabo. RCS: A reference model architecture for intelligent vehicle and highway systems. In *Proc. of the 25th Silver Jubilee International Symposium on Automotive Technology and Automation*, Florence, Italy, June 1992.

[5] J. S. Albus, H. G. McCain, and R. Lumia. NASA/NBS standard reference model for telerobot control system architecture (NASREM). *Technical Note 1235*. National Institute of Standards and Technology, Gaithersburg, MD, April 1989.

[6] J. S. Albus, C. McLean, A. J. Barbera, and M. L. Fitzgerald. An architecture for real-time sensory-interactive control of robots in a manufacturing environment. In *Proc. of the 4th AC/IFIP Symposium on Information Control Problems in Manufacturing Technology*, Gaithersburg, MD, October 1982.

[7] J. S. Albus and A. Meystel. *Engineering of Mind: An Introduction to the Science of Intelligent Systems*. John Wiley and Sons, New York, 2000.

[8] P. J. Antsaklis and K. M. Passino (eds.). *An Introduction to Intelligent and Autonomous Control*. Kluwer Academic Publishers, Norwell, MA, 1993.

[9] A. J. Barbera, J. S. Albus, and M. L. Fitzgerald. Hierarchical control of robots using microcomputers. In *Proc. of the 9th International Symposium on Industrial Robots*, Washington, DC, March 1979.

[10] A. J. Barbera, J. S. Albus, M. L. Fitzgerald, and L. S. Haynes. RCS: The NBS real-time control system. In *Proc. of the 8th Robots Conference and Exposition*, Detroit, MI, June 4–7 1984.

[11] D. Bertsekas and R. Gallager. *Data Networks*. Prentice Hall, Upper daddle River, NJ, 1992.

[12] P. Cho and J. K. Hedrick. Automotive power train modeling for control. *Journal of Dynamic Systems, Measurement, and Control*, 111:568–576, December 1989.

[13] W. Chu and M. Tomizuka. Experimental study of a lane change manuever for AHS applications. In *Proc. of the American Control Conference*, pages 139–143, 1995.

[14] W. Chu and M. Tomizuka. Lane change manuever of automobiles for the intelligent vehicle highway system. In *Proc. of the American Control Conference*, pages 3586–3587, 1997.

[15] H. M. Deitel. *Operating Systems*. Addison-Wesley, Reading, MA, 2nd edition, 1990.

[16] H. M. Deitel and P. J. Deitel. *How to Program C++*. Prentice Hall, Upper daddle River, NJ, 1994.

[17] C. J. Dixon. *Tires, Suspension, and Handling*. SAE Publishing, Warrendale, PA, 1996.

[18] J. R. Ellis. *Vehicle Dynamics*. London Business Books Ltd., London, 1969.

[19] R. Fenton, G. Melocik, and K. Olson. On the steering of automated vehicles: Theory and experiment. *IEEE Trans. on Automatic Control*, 21(3):306–315, June 1976.

[20] R. E. Fenton and I. Selim. On the optimal design of an automotive lateral controller. *IEEE Trans. on Vehicular Technology*, 37(2):108–113, May 1988.

[21] V. Gazi, M. L. Moore, and K. M. Passino. Real-time control system software for intelligent system development: Experiments and an educational program. In *Proc. of the IEEE Int. Symp. on Intelligent Control*, pages 102–107, Gaithersburg, MD, September 1998.

[22] J. Guldner, H. S. Tan, C. Chen, and S. Patwardhan. Changing lanes on automated highways with look down reference systems. In *Proc. of the IFAC Workshop on Advances in Automotive Control*, pages 69–74, 1998.

[23] C. Hatipoğlu, Ü. Özgüner, and M. Sommerville. Longitudinal headway control of autonomous vehicles. In *Proc. of the IEEE Conference on Control Applications*, pages 721–726, September 1996.

[24] C. Hatipoğlu, Ü. Özgüner, and K. Ünyelioğlu. On optimal design of a lane change controller. In *Proc. of the 1995 Intelligent Vehicles Symposium*, pages 436–441, 1995.

[25] C. Hatipoğlu, K. Redmill, and Ü. Özgüner. Steering and lane change: A working system. In *Proc. of the IEEE Conference on Intelligent Transportation Systems*, pages 272–277, November 1997.

[26] T. Hessburg, H. Peng, and M. Tomizuka. An experimental study on lateral control of a vehicle. In *Proc. of the American Control Conference*, pages 3084–3089, 1991.

[27] A. Hsu, F. Eskafi, S. Sachs, and P. Varaiya. Protocol design for an automated highway system. *Discrete Event Systems: Theory and Applications*, 2(3/4):183–206, 1993.

[28] H. M. Huang. Hierarchical real-time control task decomposition for a coal mining automation project. *NISTIR 90-4271*. National Institute of Standards and Technology, Gaithersburg, MD, 1990.

[29] H. M. Huang. An architecture and methodology for intelligent control. *IEEE Expert*, 11(2):46–55, April 1996.

[30] H. M. Huang, J. S. Albus, W. Shackleford, H. Scott, T. Kramer, E. Massina, and F. Proctor. An architecting tool for large-scale system control with an application to a manufacturing workstation. In *Proc. of the 4th International Software Arhitecture Workshop, in Conjuction with the 22nd International Conference on Software Engineering*, Limerick, Ireland, June 2000.

[31] H. M. Huang, R. Hira, and R. Quintero. A submarine maneuvering system demonstration based on the NIST real-time control system reference model. In *Proc. of the 8th IEEE Int. Symp. on Intelligent Control*, Chicago, September 25–27 1993.

[32] H. M. Huang, R. Quintero, and J. S. Albus. A Reference Model, Design Approach, and Development Illustration Toward Hierarchical Real-Time System Control for Coal Mining Operations volume 46 of *Control and Dynamic Systems Book Series* (C. T. Leondes, Ed.), chapter in *Manufacturing and Automation Systems: Techniques and Technologies*, part 2 of 5, pages 173–254. Academic Press, San Diego, CA, 1991.

[33] H. M. Huang, H. Scott, E. Massina, M. Juberts, and R. Quintero. Intelligent System Control: A Unified Approach and Applications chapter in *Expert Systems Techniques and Applications* (C. T. Leondes, Ed.). Academic Press, San Diego, CA, 2000.

[34] S. Kato, K. Tomita, and S. Tsugawa. Lane change manuevers for vision based vehicle. In *Proc. of the IEEE Conference on Intelligent Transportation Systems*, pages 129–134, 1997.

[35] E. W. Kent and J. S. Albus. Servoed world models as interfaces between robot control systems and sensory data. *Robotica*, 2(1), January 1984.

[36] D. J. McMahon, J. K. Hedrick, and S. E. Schladover. Vehicle modeling and control for automated highway systems. In *Proc. of the American Control Conference*, pages 297–303, 1990.

[37] M. Milenkovic. *Operating Systems: Concepts and Design.* McGraw–Hill, New York, 1987.

[38] B. Overland. *C++ in Plain English.* MIS Press, New York, 1996.

[39] Ü. Özgüner, C. Hatipoğlu, A. İftar, and K. Redmill. Hybrid Control Design for a Three Vehicle Scenario Demonstration Using Overlapping Decompositions *Hybrid Systems*, chapter IV, pages 294–328. Springer-Verlag, New York, 1997.

[40] Ü. Özgüner, C. Hatipoğlu, and K. Redmill. Autonomy in a restricted world. In *Proc. of the First IEEE ITS Conference*, page 283, Boston, November 9–12 1997.

[41] H. Peng. A theoretical and experimental study on vehicle lateral control. In *Proc. of the American Control Conference*, pages 1738–1742, 1992.

[42] H. Peng and M. Tomizuka. Vehicle lateral control for highway automation. In *Proc. of the American Control Conference*, pages 788–794, 1990.

[43] R. Quintero and A. J. Barbera. A software template approach to building complex large-scale intelligent control systems. In *Proc. of the 8th IEEE Int. Symp. on Intelligent Control*, pages 58–63, Chicago, September 25–27 1993.

[44] R. Quintero and A. J. Barbera. A task oriented approach for developing complex large-scale intelligent control systems using software templates. In *Proc. of the 8th IEEE Int. Symp. on Intelligent Control*, Chicago, September 25–27 1993.

[45] K. Redmill and Ü. Özgüner. The Ohio State University automated highway system demonstration vehicle. In *Proc. of the International Congress and Exposition*, Detroit, MI, February 23–26 1998.

[46] W. Shackleford. Real-time control systems library: Software and documentation. See http://www.isd.mel.nist.gov/projects/rcslib.

[47] W. Sienel and J. Ackermann. Automatic steering of vehicles with reference angular velocity feedback. In *Proc. of the American Control Conference*, pages 1957–1958, 1994.

[48] J. A. Simpson, R. J. Hocken, and J. S. Albus. The automated manufacturing research facility of the National Bureau of Standards. *Journal of Manufacturing Systems*, 1(1), 1983.

[49] M. Sommerville, K. Redmill, and Ü. Özgüner. A multi-level automotive speed control. In *Proc. of the 1996 SAE International Congress and Exposition*, February 1996. SAE Paper 960515.

[50] J. D. Spragins. *Telecommunications, Protocols, and Design.* Addison–Wesley, Reading, MA, 1987.

[51] R. M. Stallman. *GNU Make: A Program for Directing Recompilation.* Free Software Foundation, Boston, 1996.

[52] R. M. Stallman. *Using and Porting GNU CC.* Free Software Foundation, Boston, 1996.

[53] B. Stroustrup. *The C++ Programming Language.* Addison-Wesley, Reading, MA, 3rd edition, 1997.

[54] S. Szabo, H. Scott, K. Murphy, and S. Legowik. Control system architecture for remotely operated unmanned land vehicle. In *Proc. of 5th IEEE Int. Symp. on Intelligent Control*, Philadelphia, September 5–7 1991.

[55] K. Ünyelioğlu, C. Hatipoğlu, and Ü. Özgüner. Design and stability analysis of a lane following controller. *IEEE Trans. on Control Systems Technology*, 5(1):127–134, January 1997.

[56] K. Valavanis and G. Saridis. *Intelligent Robotic Systems: Theory, Design, and Applications.* Kluwer Academic Publishers, Norwell, MA, 1992.

[57] P. Varaiya. Smart cars on smart roads: Problems of control. *IEEE Trans. on Automatic Control*, 38(2):195–207, February 1993.

[58] J. Zumberge and K. M. Passino. A case study in intelligent control for a process control experiment. In *Proc. of the IEEE Int. Symp. on Intelligent Control*, pages 37–42, Dearborn, MI, 1996.

[59] J. Zumberge and K. M. Passino. A case study in intelligent vs. conventional control for a process control experiment. *Journal of Control Engineering Practice*, 6(9):1055–1075, 1998.

Index